SwidPowell
Objects By Architects

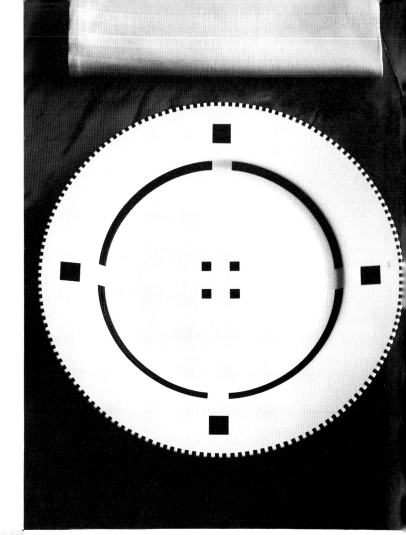

13424 BIFMC

BELFAST INSTITUTE

3 7777 00029 9396

BELFAST PUBLIC LIBRARIES
745.4
TAPE
138053

BELFAST PUBLIC LIBRARIES

DATE DUE	DATE DUE	DATE DUE
RIFHE		

13424 BIFMC

BELFAST PUBLIC LIBRARIES
745.4
TAPE
138053

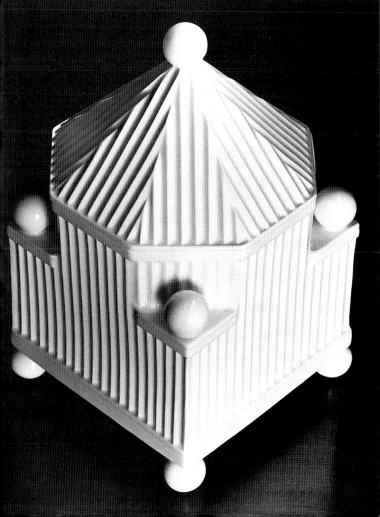

First published in Great Britain in 1990 by
Thames and Hudson Ltd., London

Originally published in the United States of America in 1990 by
Rizzoli International Publications, Inc.
300 Park Avenue South, New York, NY 10010

Copyright © 1990 Rizzoli International Publications, Inc.
and Swid Powell, Inc.
Text copyright © 1990 Annette Tapert

All Rights Reserved. No part of this
publication may be reproduced or
transmitted in any form or by any
means, electronic or mechanical,
including photocopy, recording or
any other information storage and
retrieval system, without prior
permission in writing from the
publisher.

Printed and bound by Dai Nippon, Tokyo, Japan

Annette Tapert

Objects By Architects

Introduction by Paul Goldberger

Design by Smatt Florence

B.I.F.H.E.
LIBRARY
MILLFIELD

Thames and Hudson

Table of Contents

Acknowledgments
9

Introduction
11

The Formation of Swid Powell
17

Arquitectonica
37

Michael Graves
39

Gwathmey Siegel
49

Zaha Hadid
59

Robert and Trix Haussmann
61

Steven Holl
69

Arata Isozaki
71

Richard Meier
73

David Palterer
83

Paolo Portoghesi
85

Ettore Sottsass
87

George Sowden
95

Robert A.M. Stern
97

Stanley Tigerman
107

Robert Venturi
115

The Design Process
125

Bibliography and Credits
144

Preface

One of the biggest joys in life is to have an idea blossom into a reality. In 1982 we took the plunge to combine our love for the arts with our desire to bring the artistry of world-famous architects to home product design. Our business started as a friendship and has grown into a circle of dedicated and committed architects and coworkers throughout the world. Today, our products are appreciated by consumers and are displayed in the design wings of the world's most important museums. As we reflect upon the journey which has brought us to this point, we feel blessed by this encouragement.

All of this has inspired us to devote additional energies to introducing new products with the same magical aesthetics as those that fill these pages. This book is a celebration— and provides us with an opportunity to say "thank you."

Nan Swid, Addie Powell
July 1990

Pages 2–3: *Notebook* cup and saucer by Robert Venturi; *Tuxedo* plate by Gwathmey Siegel; *Cookie Jar* by Stanley Tigerman; *Bowl* by Richard Meier.
Page 6: The Swid Powell showroom, New York City.
Page 10: Models, prototypes and production samples fill the shelves in Swid Powell's design and development office.
Page 13: Candlesticks designed by New York architect Steven Holl.
Pages 14–15: Nan Swid (left) and Addie Powell, the founders of the firm specialized in the production of objects designed by the most renowned architects and designers of the late twentieth century.

Acknowledgments

We would like to thank the following people for their generous support, help and time during the last seven years. All have made a lasting contribution to the Swid Powell company.

Architects and designers: Arquitectonica, Javier Bellosillo, Joseph D'Urso, Peter Eisenman, Frank Gehry, Michael Graves, Gwathmey Siegel, Zaha Hadid, Robert and Trix Haussmann, Steven Holl, Hans Hollein, Arata Isozaki, Philip Johnson, Donald Kaufman, Robert Mapplethorpe, Richard Meier, David Palterer, Charles Pfister, James Stewart Polshek, Paolo Portoghesi, Andrée Puttman, Fred Schwartz, SITE, Ettore Sottsass, George Sowden, Robert A.M. Stern, Donald Sultan, Matteo Thun, Stanley Tigerman and Margaret McCurry, and Robert Venturi and Denise Scott Brown.

Family: Bradford and Adam Powell, Stephen, Robin, Scott and Jill Swid.

Friends: Arnell/Bickford, Eric Boman, Franz Brahweiles, Adrienne Catropa, Michael Duncan, Michael Favitta, Joseph Feminella, Bill Georgis, Joel Goldstein, Yohiro Gotoh, Lisa Green, Amy Grossman, Robert Isabel, Janet Kares, Heidi Kay, Linda Kinsey, Stephen Knoll, Denham Lent, Harold and Lynn Levin, Nancy Lopez, Ted Marlow, Frank McIntosh, Guido Mayer, R. Craig Miller, Alessandro Munari, Cleto Munari, Maasaki Norita, Taco Oldenberger, Andrea Piccioli, Tony Rizzotti, Michael Roberts, Fred Rosen & Associates, Edmund Schaffzin, Gary Schonwald, Ian Schrager, Stephen Sills and James Hunniford, Skolos, Wedell and Raynor, Smatt Florence Inc, Dieter Spichal, Donald Strum, Geraldine Stutz, Wim Tijssen, Julian Tomchin, George Trescher, Albert Watson and Maurice Weintraub.

Stores: Bergdorf Goodman, Bloomingdale's, Bullock's, By Design, D.F. Sanders, Marshall Field's, Neiman-Marcus, Table of Contents and The American Hand.

Swid Powell staff: Carina Courtright, Marc Hacker, Dorte Hansen, Nicole Karaman, Rudolph McKinnley, Jan Newell, Tracy Schaffzin, Johanna Semple, James Stubbs, Edward Thomas, Jr. and Robert de la Torre.

Writers and editors: Charles Gandee, Paul Goldberger, Arlene Hirst, Robert Janjigian, Jesse Kornbluth, Annette Tapert and Carol Vogel.

Introduction

Paul Goldberger

It is a curious phrase, "the tabletop industry": it seems both quaint and overbearing. The very words make it clear why Swid Powell filled a void. Nan Swid and Addie Powell had little interest in the sentimental, sweet prettification of the household object that most manufacturers of plates and glassware have traditionally made their mission, and they had still less interest in thinking of themselves as an industry. Their goal was much simpler, and much more daring: to commission items that suited their own tastes, and to bring these things successfully to the marketplace.

It is difficult, now that Swid Powell has become virtually a standard brand name in certain circles, to remember just how radical an idea this was at the beginning. A decade ago, the commercial output of plate and glassware makers seemed fairly equally divided between the banal and the trite: plates that looked like the wallpaper in a coffee shop, glassware that looked like rock-cut crystals. Bonnier's, that great temple of modern design on Madison Avenue, was long gone, its name recalled by only those New Yorkers whose memories also harked back to Best's and DePinna's; so, too, with the old Georg Jensen on Fifth Avenue. People who needed to shop in the present—and who did not want fleur-de-lis on their plates or cut crystal in their wine goblets—had few places to go. (It is somewhat ironic that in those days the best outlets for small, well-designed household objects were the mass-market modern-design stores. Many architects, torn between their innate snobbism and their desire for clean, simple objects, chose Pottery Barn over Wedgwood.)

Into the breach stepped Swid and Powell, determined to bring high-quality pieces of contemporary design to the marketplace. That they knew so little about the so-called tabletop industry was, in more ways than one, their salvation: they were undaunted by the industry's own prejudices, and at the same time they were energized by the world they did know, which was the field of contemporary furniture, interior design and architecture. They saw a rising demand for work by well-known architects and designers, and they understood the extent to which these architects had begun to move into the popular realm. They had observed architects like Charles Gwathmey and Robert Siegel, Richard Meier and Robert A.M. Stern devote more and more of their practice to the design of interiors. (Even as these architects continued to seek commissions to design skyscrapers, they were increasingly eager to take on the design challenges raised by small-scale, private spaces.) To Swid and Powell, it was a logical step to the next level down in scale: as the architects had gone from buildings to rooms, so they could go from rooms to objects within the rooms.

The most important legacy of those years was the growing recognition that architecture had a sensuous dimension—the increasing willingness of architects to admit that it was possible to think of rooms, buildings, of entire cities in emotional as well as intellectual terms. Modernism had, by and large, disdained the sensuous; while there was great lushness in the work of Mies van der Rohe, there was nothing but coldness in the mediocre buildings of his imitators that filled every downtown. This, more than anything else, was the real achievement of Post-Modernism: it gave sensuousness a degree of intellectual respectability in design that it had not possessed for half a century. And this, in turn, made it possible for architects to accept the idea of designing household objects.

Swid and Powell cleverly perceived something else, too: that in the mid-1980s, architects had begun to take on the trappings of celebrity. While Philip Johnson was hardly the first architect to appear on the cover of *Time*, his presence there (as well as on the cover of *The New York Times Magazine* holding a model of the A.T. & T. Building) seemed to herald a new era—an era whose quintessence, perhaps, was in the commission Michael Graves received to do a shopping bag for Bloomingdale's. Not all was frivolous: for every architect whose comings and goings were documented in *People* or the gossip columns, there was another who attracted serious attention with a new exhibition of drawings or a museum show. The buying and selling of architectural drawings and artifacts had expanded from a minor sideshow in the art market to a major business.

There were moments when it was not entirely clear whether the public's interest in the architects themselves emerged out of a rising interest in architecture, or the other way around. Was personal celebrity the cause or the effect? Either way, the 1980s represented a startling change in the nature of a profession that had traditionally seen itself as somewhat aloof from the public, possessed of a somewhat self-conscious, and more than occasionally even self-righteous, distance from the public eye. Suddenly architecture was intersecting more closely with the consumer society, and even beginning to look more than a little bit like fashion—more popular, more visually exciting than it had been before, but also more changeable, less able to withstand the shifting tides of popular taste.

Swid and Powell saw the potential in such a moment—and, to their credit, they saw the dangers. It would have been easy to have exploited the newfound identity of architects as kin to fashion designers; Swid Powell could probably have sold toothbrushes with some architects' signatures on them. To Swid and Powell's credit, they never wavered from their original intention—and in fact found themselves rejecting design proposals from several architects, including some of the most celebrated in their stable, for objects that they did not believe met their standard.

Now, six years after the first set of Swid Powell plates and five years after the company's line expanded to include a variety of other objects, there is a significant body of work to evaluate. What is clear is that Swid Powell represents no particular design ideology, no narrow stylistic approach, but seems committed to commissioning the best work from a wide range of designers. The quality of craftsmanship is consistently high; there is an elegance of finish as well as of conception to all of the pieces. But most striking is the extent to which these pieces have turned out to be a kind of small-scale laboratory for their designers, enriching the nature of their œuvre by giving them a chance to explore design issues in a small, tightly controlled situation. Several surprises occur: Gwathmey Siegel, in their *Tuxedo* plate (the single best-selling item Swid Powell has produced, and by now a kind of icon for the entire company) managed to make the predisposition toward geometric order that is central to their architecture the basis for an elegant, even lush object. Meier was less inventive than Gwathmey Siegel in the design of his plates, but his bowls, trays, candlesticks and desk implements, many of which indicate a more than passing debt to the work of Josef Hoffmann, together make up a significant collection in themselves; here, too, geometric precision yields splendid luxury. As might be expected, Stern has come close to a literal recall of historical objects with his *Moderne* and *Swag* plates, while Robert Venturi avoids the literal past but manages to be even more powerfully evocative with his *Notebook* plate, based on the white-on-black abstract design of old student composition books, and his *Grandmother* pattern, a kind of mix of pastel flowers and Jasper Johns crosshatching.

Although there are things that could be said about any of the other Swid Powell pieces—the Tigerman McCurry tea set and salt and pepper shakers, a witty china version of the architects' country house; the gently classicizing Michael Graves *Corinth* plates; and the intensely colored George Sowden *Rio* plates are probably the best of the objects not yet mentioned—in the end, it is the collection as a whole that is the thing. The Swid Powell objects forge a new point of intersection between architecture and the realm of consumer products—a point of intersection that could not have existed a few years ago. They find this point not at the lowest common denominator of fashion, but at a higher, more serious level—not by bringing architecture down to the level of commercial products, but by bringing commercial products up to the level of architecture.

The Formation of Swid Powell

The early 1980s produced a curious disparity in the home-furnishings market. Despite a growing interest in American architecture, a heightened appreciation of interior design and a revival of the decorative arts, the home-accessories industry was moribund. People looking for china almost invariably thought of the traditional five-piece place setting, and when it was time to buy silver consumers were still buying the heavy, traditional English pieces that their parents had.

For Nan Swid, a former design director in product development at Knoll International, this lack of choice was unsettling. "I realized that beautiful things for contemporary homes were missing from the market," she recalls. "I found I was buying fine old things when I am really a Modernist at heart. What I wanted was a portrait of my generation. I also knew that very few people would ever live in houses designed by Richard Meier or Robert Venturi, but I thought they would like to experience that aesthetic level."

Opposite: Entry door plaque at Swid Powell headquarters, 213 East 49th Street, New York City. Right: (left to right) Peter Eisenman, Michael Graves, Nan Swid, Charles Gwathmey, Stanley Tigerman, Addie Powell, Robert Siegel and Robert A. M. Stern.

Powered by that single perception, Swid envisioned a business that would have leading contemporary architects design a line of functional but beautiful tabletop accessories for the home—china, crystal, silver. In 1982, she approached Addie Powell, a former vice president in sales at Knoll, and proposed that they fill what looked like a permanent market void.

In November 1982, Swid and Powell invited nine architects to lunch at the most fitting restaurant in the world for a meeting of this kind—the Four Seasons in New York, designed by Philip Johnson. Some of these architects had worked with Swid when designing furniture at Knoll; all were her friends. Philip Johnson, Richard Meier, Stanley Tigerman, Charles Gwathmey, Robert Siegel, Robert A.M. Stern, James Stewart Polshek, Charles Pfister and Joe D'Urso listened as Swid and Powell shared their hopes for the new venture.

"They knew what we were going to do," explains Swid, "but we wanted to introduce them to the business end—how we were going to manufacture the items, how they would be marketed, what kinds of ideas we had. Though we didn't have answers to all the questions, we wanted it to be taken as a serious business."

One architect stood to toast the venture. Another said, "Sit down, we've got too much work to do." Despite that commitment, there was still some skepticism. "There were some who left the restaurant that day very doubtful that Nan and Addie's venture could ever happen," Gwathmey recalls. "I disagreed. I told the skeptics it could be done if we were all philosophically and pyschologically supportive. We had to join in the system."

Given a few directions—Swid and Powell listed the materials that could be used in the product categories—the architects promised to return the following January with sketches or models of their designs. At that meeting, they agreed, they would critique one another's work. Swid and Powell hired a photographer to immortalize the event; they knew it would be historic.

The January meeting produced, as Siegel recalls, "a series of somewhat bizarre designs that would have been enormously expensive to make and probably would have ended up as one-off museum pieces." Nevertheless, the architects had done their homework.

Meier appeared with a four-foot balsa-wood model of a candelabra that, says Powell, "could have gone into the Vatican." Gwathmey Siegel delivered a model of a bowl that would, they hoped, be made in crystal. "It was so geometrically driven that even our cabinetmaker couldn't make it in plastic, and it

Left: Tigerman discusses his design sketches at an initial design review, 1983. Right: (left to right) Tigerman, Charles Pfister, Powell, Gwathmey, James Stewart Polshek, Meier and Stern evaluate a proposed design by Meier at a design review.

Overleaf: Swid and Powell share an office and desk at the firm's headquarters, formerly the apartment/office of decorator Billy Baldwin. Page 21: Swid, Graves, Stern and Tigerman at the opening party of the firm's headquarters, 1988.

cost us $1000 to make a paper model because he couldn't put it together. Not only couldn't you make a mold for it, you couldn't have gotten it out of the mold," Gwathmey remembers. Tigerman arrived with a candelabra that used four different kinds of material—a manufacturing impossibility. "At that time we hadn't even learned to work in one material," says Swid.

For Swid and Powell, there was genuine concern about putting all these architects in one room for a design critique. "We were worried they'd be too brutal and it would end in disaster," says Powell. But as Tigerman explains, "Most of us were really good friends. Some of us were at Yale at the same time, over the years we've been on juries together and we've competed with one another for commissions. We can be ruthless and we can be quite critical of one another's work, but there's an authentic affection amongst us. We were tough and utterly honest at those early meetings, but we also made useful suggestions. And, in a certain way, we care what our colleagues think."

No matter that this presentation produced designs that might never be executed. It left the group energized and ready for the next presentation. At each of the following meetings, the architects would return with something different or a variation on their

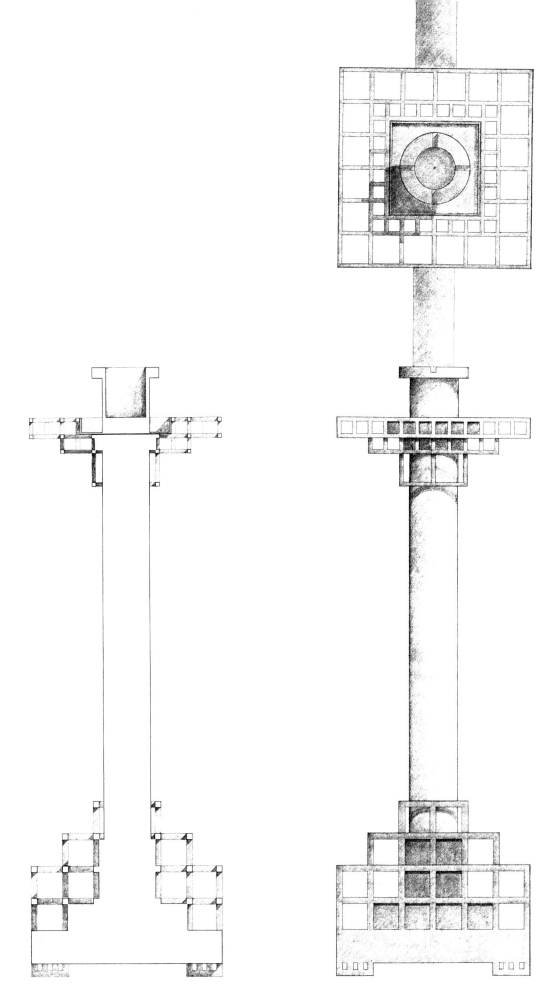

Opposite: Section, elevation and plan for Meier candlestick design, presented at initial design review in 1983. The object was deemed too complicated to manufacture.
Left: Swid and Meier discuss new designs at the architect's New York office.
Below: Section, elevation and plan for Meier *Silver Rose* bowl, also not produced.

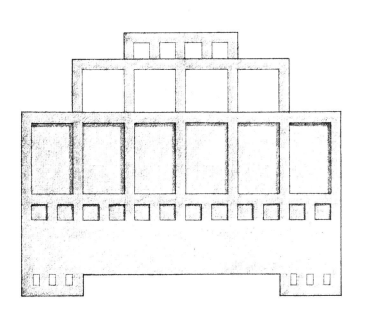

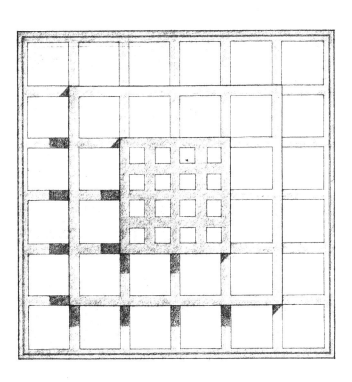

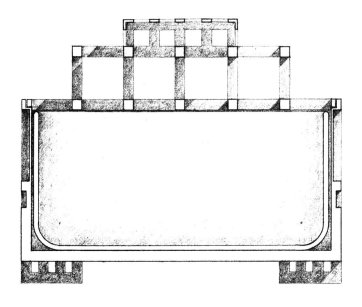

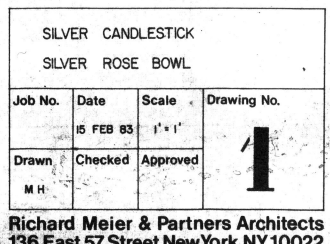

original designs. "Those were fun days, and in some ways our designs were highly autobiographical," recalls Tigerman. "There was one architect who was sort of doing his Gucci loafers on his dinner plates."

What Swid remembers most about those meetings was how supportive everyone was. "If one architect arrived unprepared, the others got very perturbed. They genuinely wanted to get this project off the ground," she says. "Addie and I had done a study on the tabletop industry, but we had no experience in manufacturing and retailing. It was out of friendship that the first group of architects went out of their way to help us. It was a risk, and it took an enormous amount of their time and effort, but they plunged in with us."

For the architects, it was an opportunity to reestablish a link between architecture and decorative arts. Historically, architects had once been responsible for more than buildings, but not since the days of the Wiener Werkstätte, founded by Josef Hoffmann in 1903, had there been such a studio. "I liked Nan and Addie's enthusiasm, their naive way of saying let's make something and find out how to do it. If it weren't for their perseverance and dedication to making it work, it would have fallen apart," says Meier.

But what really captivated the architects was that Swid and Powell wanted to make the products accessible to the public. "For another company, I had done a silver tea service that was so prohibitively expensive I didn't own one myself," notes Meier. "It had saddened me to put all that energy into an object that would never be shared."

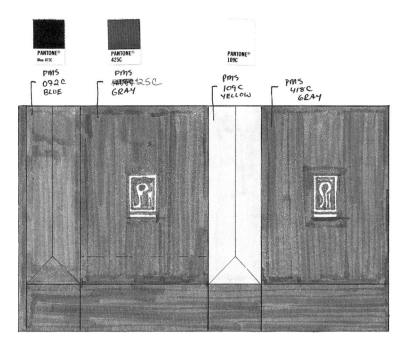

Opposite: Swid Powell shopping bags.
Top: Color plan of Swid Powell shopping bag.
Above: Image brochure, with photographs by Albert Watson, was designed by the New York firm Smatt Florence Inc.

Left: The fireplace mantel in the office occupied by Marc Hacker, Vice President of Design and Development, is a place for the display of prototypes for metal and silver-plated items, some of which are in production.

Through the meetings, Swid and Powell began to see more clearly what products seemed feasible from a manufacturing viewpoint, and the architects began to find their focus. "After the third or fourth meeting, Nan and I began to concentrate on getting the products made. We didn't concentrate on the design; we left that to the architects," says Powell.

With the help of many friends in retailing and in the design field, Swid and Powell were directed to Italian silversmiths, Austrian crystal manufacturers and Japanese porcelain makers. And then Marc Hacker, a young architect formerly with Meier's office, joined them as a liason between the architects and the factories.

In June 1984, after almost eighteen months of numerous models and prototypes, redesigning and refining, Swid and Powell presented their inaugural collection to retail buyers around the country. It was relatively small: fifty-three pieces of porcelain dinnerware, silverplate pieces and crystal stemware designed by Gwathmey Siegel, Meier, Stern, Laurinda Spear, Arata Isozaki, Tigerman and Venturi. "We selected our retailers as carefully as we selected our architects," says Powell. "Each store bought the entire collection, with the promise that they would devote a section of the store to us and give an opening party."

In November of the same year, the Swid Powell Collection was unveiled to the public for the first time at Marshall Field's in Chicago. The china sold briskly, and all the silver pieces were sold out within four days. "It has been a long time since the tabletop industry has had this kind of excitement," Marshall Field's buyer Robert Doerr remarked at the time. With some of the architects in attendance, and stores placing full-page ads with headlines like "Walk away with a Richard Meier original for considerably less than his normal fee," Swid and Powell launched their collection in six major American cities. By Christmas their entire first production order had sold out, and there were many orders from customers who were willing to wait several months for delivery.

"We were as surprised as anyone," Swid says. "While we were putting the collection together, Addie and I didn't focus on whether it would sell. Our concern was to make the best possible products, and to do that we simply chose designs we liked by architects we respected."

Eight years later, Swid and Powell work out of an office in Manhattan's Turtle Bay that was once the apartment of Billy Baldwin, another great force in design. Many of the prototypes of the original pieces stand proudly as part of the permanent design collections of the Museum of Modern Art and the Metropolitan Museum of Art in New York. "Swid Powell marked the return of American architects to the field of applied decorative arts," notes R. Craig Miller, the

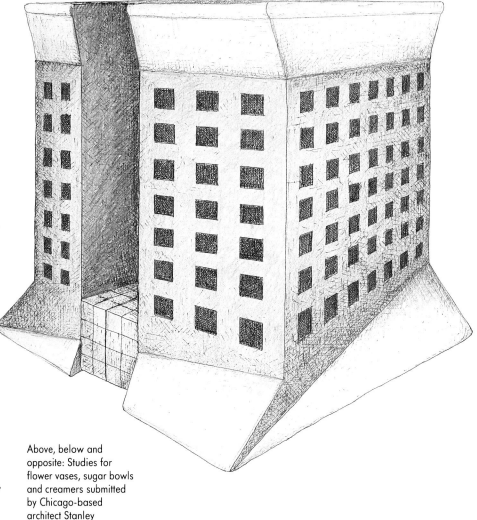

Above, below and opposite: Studies for flower vases, sugar bowls and creamers submitted by Chicago-based architect Stanley Tigerman at early design reviews. These objects were never developed further.

29

Metropolitan's twentieth-century-design curator. "We hadn't seen this since the mid-sixties, when Charles Eames and the Cranbrook circle were designing everything from rugs to tea sets to furniture. Swid Powell is the first company to design small objects at a high level of quality. Nan and Addie went out into uncharted waters. What they do has great focus."

The Swid Powell Collection now extends to approximately two hundred pieces. Many of the original designs are still in production—Meier's silver bowl, Stern's *Harmonie* candlesticks, Venturi's *Grandmother* and *Notebook* plates and Gwathmey Siegel's *Tuxedo* dinnerware, to name a few. In some instances, such as *Tuxedo*, the line has been extended into a wider range of serving pieces. And to join these "classics" are pieces designed by such architects as Michael Graves, Robert and Trix Haussman and Ettore Sottsass. The quality, the focus and the mission continue.

None of this would have been possible without the support of retailers. "It's one thing to develop a product line of fifty-three pieces, but that's meaningless without retailers who have enough vision to know that people would buy this type of product," says Powell. "Our goal was to reach upper-end department and specialty stores." Swid Powell certainly has accomplished that; its products are sold in five hundred American stores, and are distributed throughout Europe, Canada and Japan.

"Without knowing it, these two women made one of the most important contributions to the tabletop industry at a time when the business was in the doldrums," says Julian Tomchin, who first saw the collection when he was the fashion director for home furnishings at Bloomingdale's. "Before Swid Powell, it was an industry based on five-piece place settings. For example, by originally offering plates in only one size, they made it possible to sell individual items as gifts, not based on color or pattern but fueled by that rarest of commodities—an idea."

"Seven years later, re-reading the outline that we gave the architects, what we achieved was exactly what we set out to do," says Powell. "We never wanted the collection to be an elitist product. We didn't want it to be perceived only as art; we did want it to be functional. We often tell our customers not to be intimidated, to put pasta on a Robert Venturi plate, or fruit in a Richard Meier bowl, or serve tea in Stanley Tigerman's *Teaside*. Our intention was not to create a series of collectibles. For us, use brings the products to life."

Preceding pages: The dinner table in Nan Swid's New York apartment is set with a variety of objects from the Swid Powell line. Architects were responsible for the design of everything on the table except the flatware. Left and opposite: To coincide with the 1988 American Institute of Architects convention in New York City, the display windows of the Bergdorf Goodman department store were designed by architects, many of whom incorporated Swid Powell pieces into their schemes. Architects were paired with fashion designers: Robert Venturi and Christian Lacroix; Michael Graves and Giorgio Armani; Calvin Klein and Richard Meier; and Robert A. M. Stern and Bill Blass. Overleaf: A residential atmosphere pervades the showroom area of the Swid Powell office, designed by Stephen Sills of New York.

Arquitectonica

As a rule, young architects start small and build slowly. For the Miami-based architectural firm of Arquitectonica, that rule couldn't be less applicable. Founded in 1977 by Laurinda Spear and her husband, Bernardo Fort-Brescia, the firm launched itself with a shocking-pink house for Spear's parents. With astonishing speed, the couple went on to build a trio of elaborate high-rise condominiums along Miami's Brickell Avenue that forever changed the city's lusterless skyline. One of them, the Atlantis, was seen by millions of Americans every Friday night for five years in the opening credits of the television series "Miami Vice."

Since their meteoric rise, these brash Modernists have gone as far as Peru, where they planned the Banco di State Creditale in Lima and up the Eastern seaboard to Herndon, Virginia, where they designed the Center for Innovative Technology.

Whether the project is a skyscraper or a plate, the problem for Spear always boils down to the same thing. "What I've learned is that working on a small scale is the same as working on a building. I still have an 8½-by-11-inch piece of paper in front of me with a design problem that I have to abstract," she explains. Her *Miami Beach* buffet plate was the solution to one such problem. "It was meant to be a service plate as opposed to a regular plate for eating. On a regular plate, the decoration tends to be around the edge. Traditionally, the pattern covers the entire surface of a service plate, so the *Miami Beach* plate has, more or less, abstract food all over it."

Opposite: Taggart town houses, Houston, Texas, 1985.
Right: The *Miami Beach* buffet plate.

Michael Graves

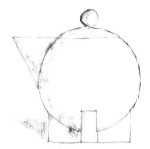 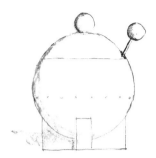 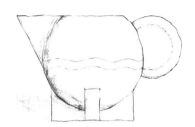

"Michael Graves's architecture more closely approximates a literary art than any other architecture one can think of in history," says Yale art historian Vincent Scully. It certainly has drama and conflict. Until the 1970s, Graves was a highly successful Modernist. Then he seemed to turn his back on Modernism in order to incorporate elements not seen since the days of classical architecture. A champion of planes, columns and geometric space, Graves had suddenly fallen in love with ornamentation, illusion and nostalgia.

In the past decade, Graves has confounded Americans with buildings that take his revelations of the 1970s even further. His Portland Building—a massive and multicolored structure that uses his signature keystone motif to suggest a wealth of literary and narrative references—has come to be regarded as an icon of Post-Modernism. He has also caused a stir with Louisville's Humana headquarters, a winery in California's Napa Valley and, most recently, the Swan and Dolphin

Opposite: *The Little Dripper* coffeepot, a ten-cup drip-filter design.
Above: Graves's sketch studies for creamers and sugars.

hotels at Walt Disney World in Florida.

"In my architectural work, my effort has been to locate myself within the body of society, so that what I do for myself is also appropriate for many others," Graves explains. "I don't see myself as breaking new ground or being terribly innovative. I try to locate my designs within the language that society understands and has been part of for many, many years. Some of my colleagues say there is no such language, but I think there is and that's what makes a building a building or a plate a plate."

For Graves, the correlation between his architecture and the pieces he has designed for Swid Powell is in the use of figurative motif and the application of color. "In many of these artifacts, I've used a similar palette, but in the symbolic sense," he notes. "For instance, on the *Little Dripper* coffee pot, the terra-cotta base represents the heat and color of coffee and the blue signifies water. This is a direct association to some extent, but with an abstract touch. In the end, though, it's ornamental, and the associations are the relationship of color and form."

But the continuity between Graves's architecture and the Swid Powell pieces is perhaps less significant than his straightforward love of products. "Everything I design is a direct result of what I want in my own home. I designed *The Little Dripper* because I'm partial to filter-drip coffee and I was dissatisfied with the aesthetics of the models that were on the market."

And then there is Graves, the formidable collector. "I have a collection of early Wedgwood plates, and I've long been fascinated by the

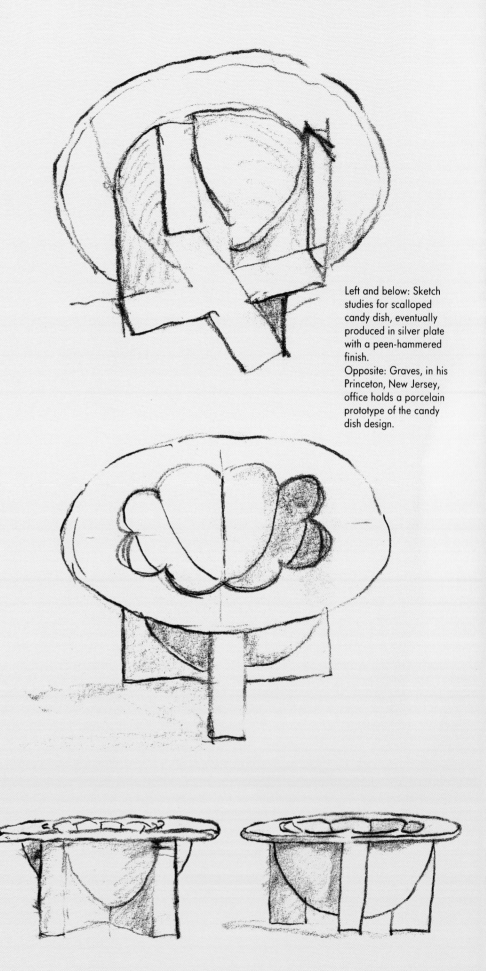

Left and below: Sketch studies for scalloped candy dish, eventually produced in silver plate with a peen-hammered finish.
Opposite: Graves, in his Princeton, New Jersey, office holds a porcelain prototype of the candy dish design.

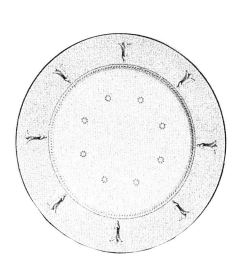

floral patterns on everyday lunch plates and dinner plates. For my Swid Powell dinnerware collections—*Delos* and *Corinth*—I thought it would be interesting to counter what the other architects might be doing. It would open up a new direction for Swid Powell. By virtue of using the palmetto—the original floral motif used in buildings, furniture and artifacts—I created a classical, floral pattern with a contemporary edge that would be suitable for Swid Powell."

As a draftsman, Graves has often been paid the compliment of seeing his architectural renderings framed and displayed. It's much less of a compliment for him to hear that his Swid Powell products are also displayed as collectibles. "I'm always amazed when people say they have my things on a shelf. I only think in terms of use."

Opposite: *Corinth* dinner plate.
Above: *Delos* dinner plate.
Right: Working drawing for silver-plated serving utensil with etched floral pattern, eventually produced by Reed & Barton for Swid Powell.

Far right top: Study for palmetto, or fleur-de-lis, used on the *Delos* plate.
Far right bottom: Hacker and Graves in conference at the architect's New Jersey office.

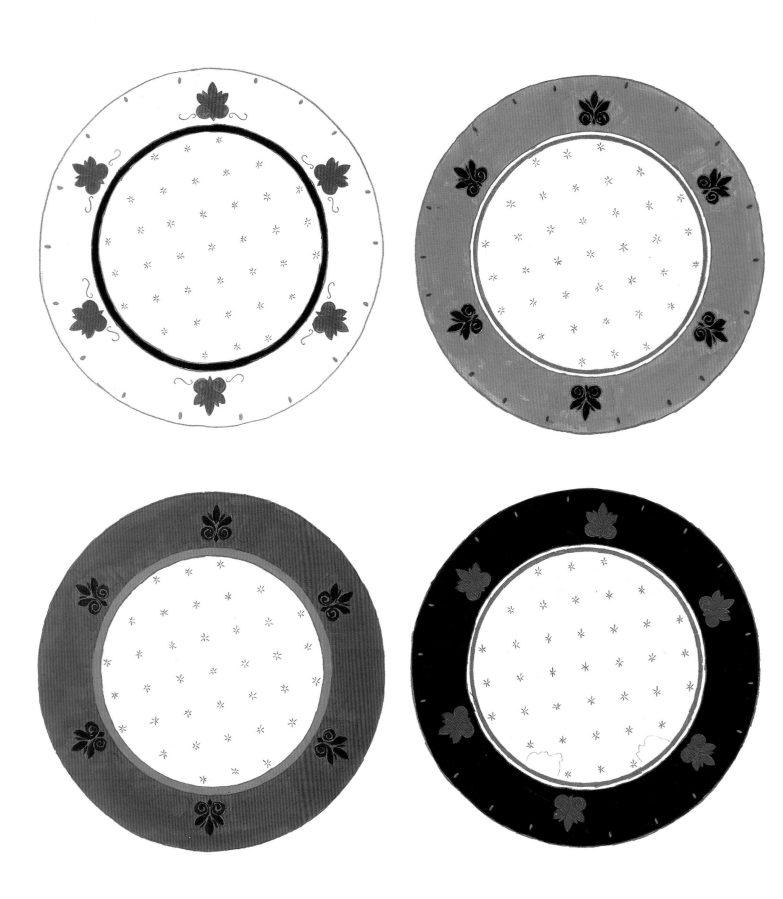

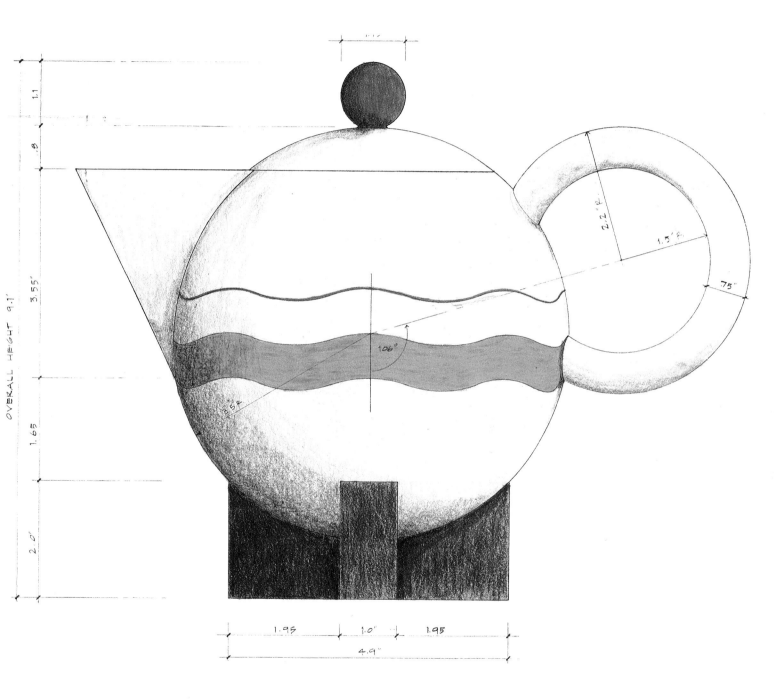

Opposite: Painted studies for *Delos* pattern.
Above: Drawing for *The Big Dripper,* a fourteen-cup version of the filter-drip coffeemaker *The Little Dripper* now produced.

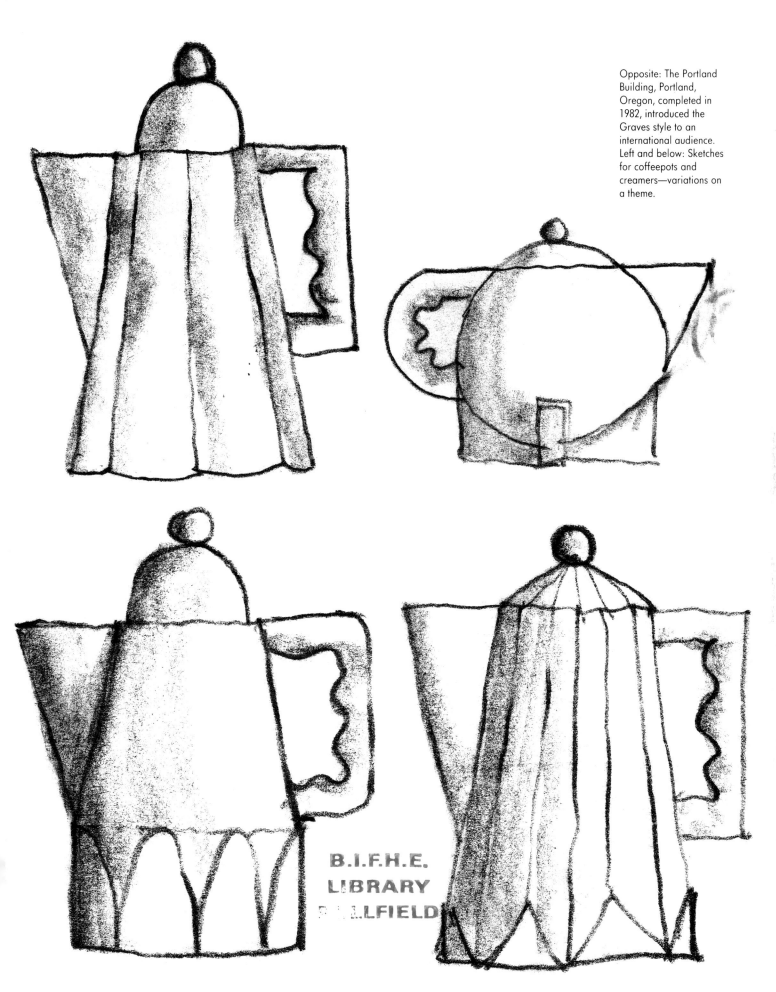

Opposite: The Portland Building, Portland, Oregon, completed in 1982, introduced the Graves style to an international audience. Left and below: Sketches for coffeepots and creamers—variations on a theme.

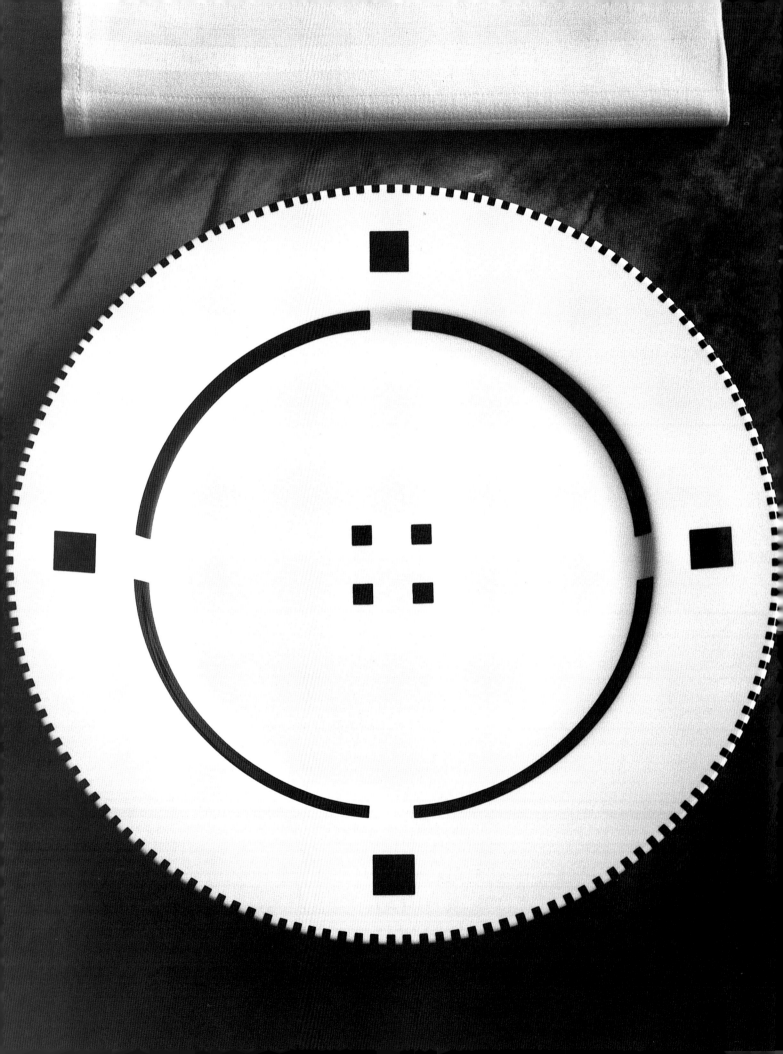

Gwathmey  Siegel

For twenty-five years, Charles Gwathmey and Robert Siegel have shared an office, a desk and some of the most provocative ideas in American architecture. Starting from a pure Modernist position, they designed houses and office buildings in which the bones of the structure were integrated into the design of the exterior. In recent years, they have taken Modernism to its next stage—softer lines, a wider palette and an increased appreciation of a site's context.

Some of the milestones of their architectural evolution include the residence and studio of Gwathmey's parents in Amagansett, New York, Whig Hall at Princeton University and the College of Architecture Building at the University of North Carolina in Charlotte. They are also creating a controversial addition to Frank Lloyd Wright's Guggenheim Museum in New York.

Gwathmey Siegel's love of clean, elegant, geometric forms has been

Opposite: The *Tuxedo* plate.
Right: Gwathmey residence (south facade), Amagansett, New York, completed 1965.

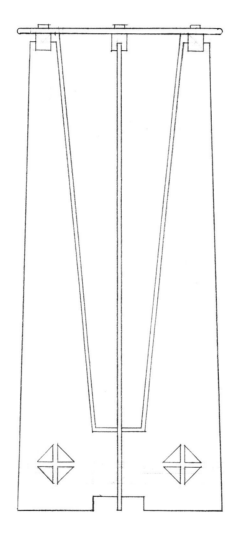
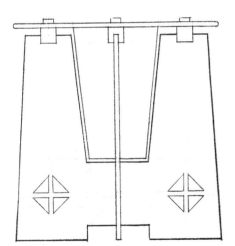

Above left: Outline drawing for *Courtney* candlestick, produced in silver plate on brass. Above: Outline drawing for *Courtney* bud vase. Left: Gwathmey and Siegel share a desk in their Manhattan office. Opposite: Gwathmey Siegel, with a display of *Tuxedo* and *Chicago* pieces and additional silver-plated items designed for Swid Powell.

translated into their designs for Swid Powell. "When Nan and Addie asked us to design a dinnerware pattern, we were very resistant to the idea of decorating a plate. It was antithetical to everything we had ever done in architecture," says Siegel. "It was a different take for us to concern ourselves with a given object, to decorate it and transform it. On the positive side, it was a good exercise. What came out of it was a very successful product—*Tuxedo*."

Both Gwathmey and Siegel realized that the formal process of designing a plate was similar to designing buildings. "What we were trying to do with *Tuxedo* was to exploit each of the formal characteristics of the object," explains Gwathmey. "We had to figure out how to articulate each of these zones in an overall design idea, but then make the design abstract, reductive and so graphically compelling that it forces the viewer to complete it. What's strong about the *Tuxedo* pattern is that it locks your eye. It's like a visual puzzle; it makes you speculate about its graphics. And that, in the end, is what makes something classical. It doesn't deal with a time frame."

And how do the products the partners have designed for Swid Powell relate to their architectural work? Gwathmey explains that they're all form-driven and analytic. "The silver vase, tray and bowl series we've done has to do with the holder and the held, the frame and the object. Again, there's the opportunity for articulating these pairs as separate objects that are joined, creating a third ingredient: how the pieces come together. So

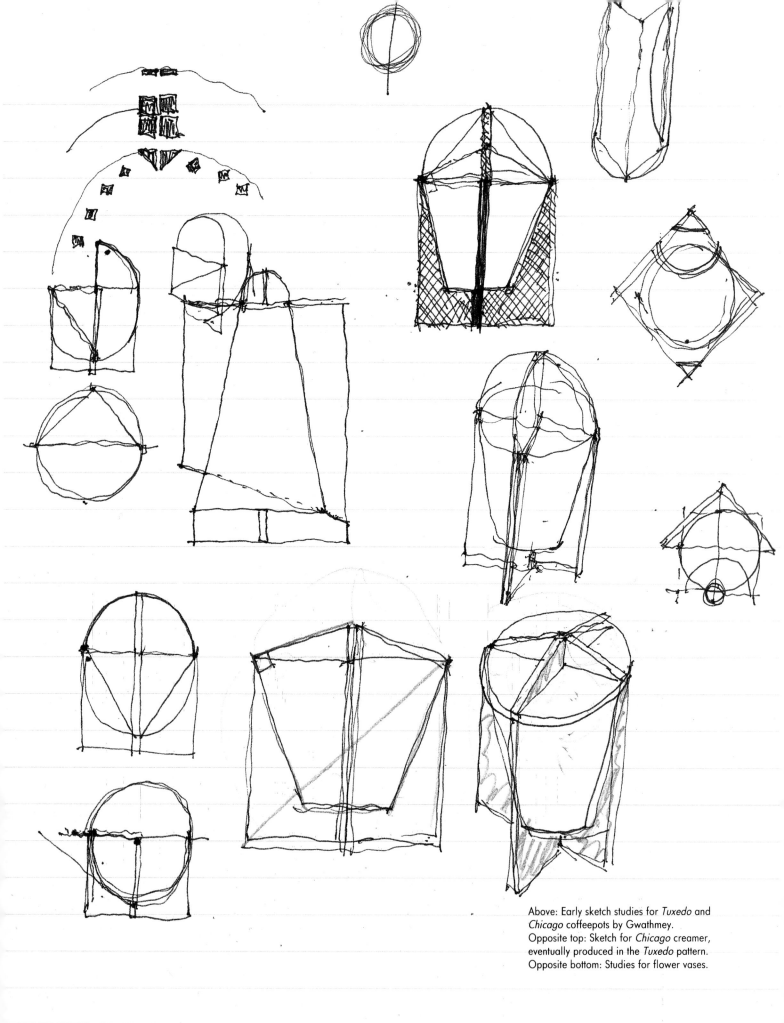

Above: Early sketch studies for *Tuxedo* and *Chicago* coffeepots by Gwathmey.
Opposite top: Sketch for *Chicago* creamer, eventually produced in the *Tuxedo* pattern.
Opposite bottom: Studies for flower vases.

the exploitation of the joint, the interconnection, the principle and form make for a construct that is articulated in the object—as it is when you make a building."

In Gwathmey and Siegel's book about their architecture, they begin the description of each site with a discussion of "the problem" and "the program" devised to solve it. They speak the same way about their Swid Powell work. "The problem-solving aspect on a small scale is the same as in architecture," says Siegel. "Once you understand the constraints of a design problem in theory, all three-dimensional problems are similar. The issues are very basic: form and composition. The difficulty in designing these objects doesn't have to do with how small they are, but with the constraints of making them."

As in their architecture, where form is derived by the demands of site, orientation and program, Gwathmey and Siegel are always thinking of use when designing for Swid Powell. "If it's not driven by what it is, if it's simply sculpture, then you shouldn't be making product," says Gwathmey. "The design strategy is to make it both compelling and useful."

"We've always believed that Modernism can still have that power and appeal. With our Swid Powell work, especially *Tuxedo*, we've seen that an abstract quality was enormously attractive to people who weren't necessarily sophisticated about that school of architecture," notes Siegel. "When something is well-designed, it doesn't have to be thought about in relation to historical periods. It has on its own a sense of repose and excitement— all the things people like."

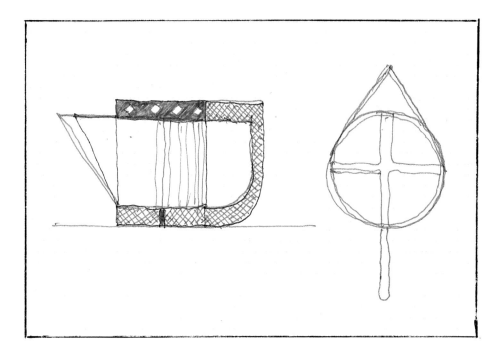

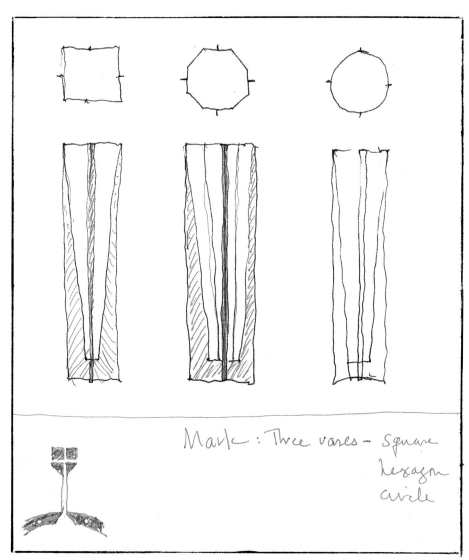

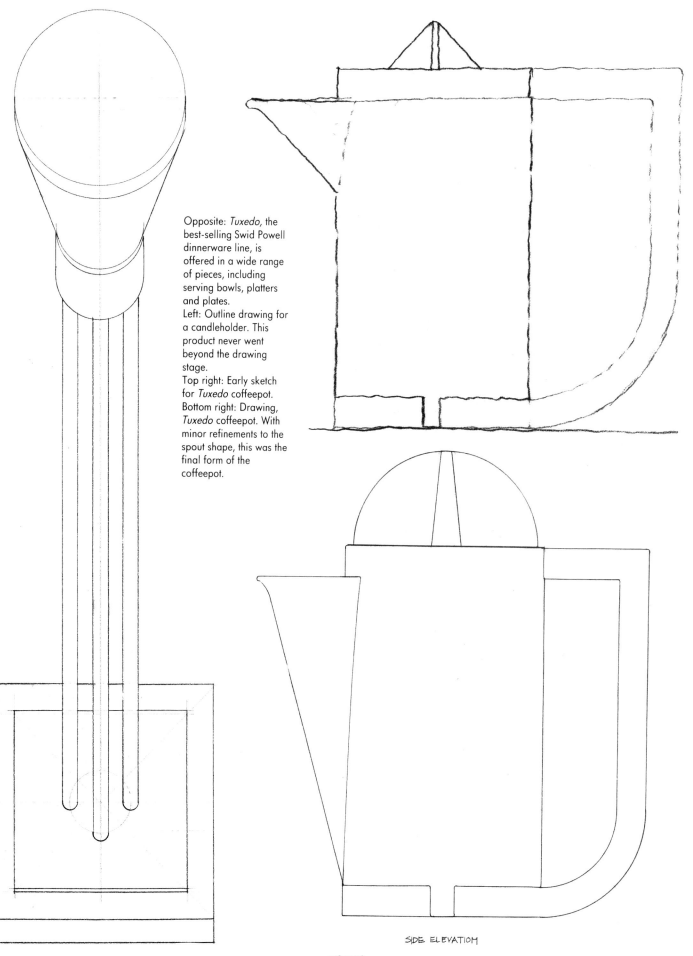

Opposite: *Tuxedo,* the best-selling Swid Powell dinnerware line, is offered in a wide range of pieces, including serving bowls, platters and plates.
Left: Outline drawing for a candleholder. This product never went beyond the drawing stage.
Top right: Early sketch for *Tuxedo* coffeepot.
Bottom right: Drawing, *Tuxedo* coffeepot. With minor refinements to the spout shape, this was the final form of the coffeepot.

SIDE ELEVATION

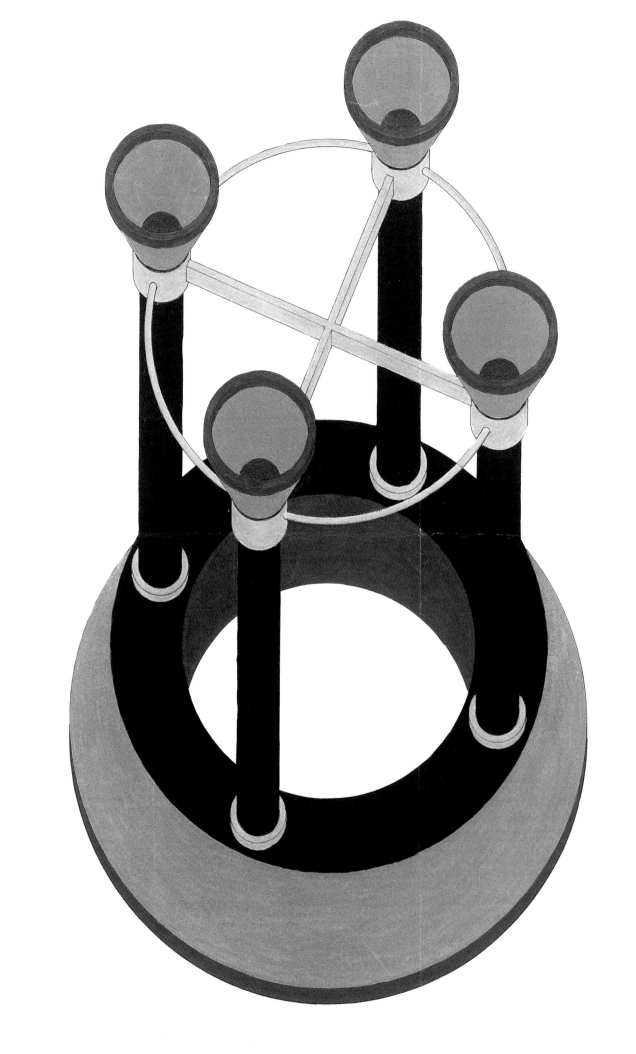

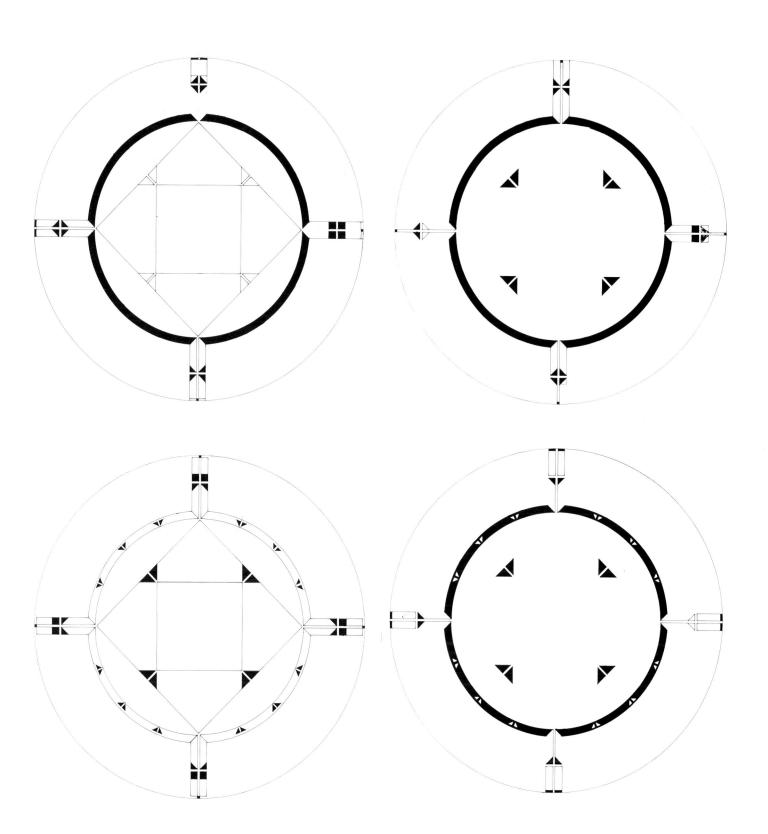

Opposite: Gwathmey Siegel's initial proposal for Swid Powell was this candelabrum.
Above: Studies for *Chicago* pattern plates. For each Swid Powell product, a number of drawings refining the pattern and its placement on the surface of the object is made throughout the development process.

Zaha Hadid

Six hundred architects entered the competition to design the Peak Club in Hong Kong in 1985. Five hundred of those entries — Zaha Hadid's among them— were immediately rejected by the jury. But competition juror Arata Isozaki rescued Hadid's entry, arguing that in the reject pile was the competition's real winner. One judge threatened to resign, but in the end, Hadid's intensely layered "horizontal skyscraper" triumphed. This unbuilt project propelled Hadid to the forefront of the contemporary architectural scene. The drawings and model of the Peak Club were the focus of her portion of the Museum of Modern Art's groundbreaking Deconstructivist Architecture exhibition of 1988.

Born in Iraq in 1950, Hadid now works in London, where she is attempting to "stabilize a new concept of Modern that would be a way of life." In addition to the Peak Club, she has designed the West Hollywood Civic Center in California, among many other projects in Germany, Japan, Switzerland and Great Britain. She has also designed a collection of furniture for the Italian firm Edra that was described by critic Cristina Morozzi as "not furnishings, but minimal entities."

One of Hadid's great strengths is draftsmanship. For her, a drawing is "a lens that reveals otherwise imperceptible aspects. It doesn't crystallize a form; it demonstrates the possibilities of what form can become." Her dinnerware for Swid Powell extends her drawing into three dimensions and reflects her interest in space, energy and the illusion of flight.

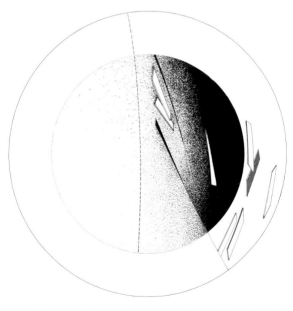

Opposite: Hadid's *Beam* pattern is applied to plates, cups, saucers and soup bowls.
Top right: Study for the energetic *Beam* pattern.
Bottom right: Decal drawing, *Beam* coffee cup.

Robert and Trix Haussmann

In their twenty-five years of working together, the Swiss husband-and-wife team of Robert and Trix Haussmann has gone from being Bauhaus architects to establishing quite a reputation as masters of the unexpected. Their style today is a result of studies they did in the early 1970s on the principles of the Classical Mannerist movement in the sixteenth century. "We tried to create our own style by expanding some of the Bauhaus ideas and taking certain aspects such as illusionism and ornamentation from our favorite period of history," says Robert Haussmann.

In Hamburg, for example, they have designed a shopping galleria with a stained-glass trompe l'oeil curtain in the display window. In the hall of the Swiss National Bank in Bern, they created isometric cubes of lacquered wood and mirrors that seem to float in space.

Opposite and above: *Stars* plate and pattern sketch.
Right: *Broken,* another pattern designed by the Haussmanns.

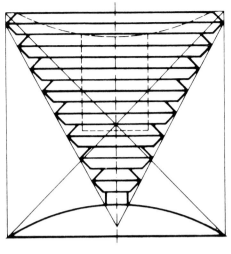

TYPE 2

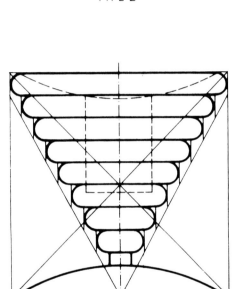

TYPE 5

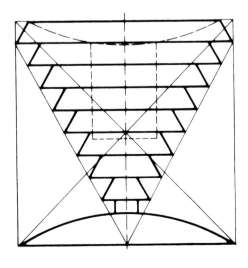

TYPE 3

And their furniture includes a wood wardrobe on which glass has been inlaid to resemble a draped cloth.

Their work for Swid Powell is equally droll. It incorporates the banal. "Stars are one of the most common—and yet meaningful—symbols, useful for both the trivial and the important," says Robert Haussmann. There is trompe l'oeil: The concept for the *Stripes* plate evolved from a piece of silk fabric they were working with in their studio. And in the *Broken* buffet plate, there's unmistakable irony. "It was a reaction against the idea of decorating a plate," say the Haussmanns. "It was a foreign concept for us not to work in form. We solved the problem with the design of a broken plate because it was, in a sense, a way of disturbing the form." Their pepper mill, with two circles as grips, is a wry take-off on Mickey Mouse's ears. But theirs is more than a Pop sensibility. Their salt and pepper shakers not only echo the silo shape of the pepper mill but when put together make a pun on the work of the well-known Italian architect Aldo Rossi, who is best known for his towers.

Working with Swid Powell has brought the Haussmanns closer to the American culture that has long interested them. It also satisfies their desire to grow as architects. "It is easier to express a pure formal idea in smaller things," they note, "because in general the small object is less complex. And we always get new ideas for architectural projects when we are thinking on a small scale."

Left: Three proposals for the Haussmann *Candlestick,* with slight differences in the shape of the stepped ridges. Above: Prototypes for a series of candlesticks in silver plate, of which only one was fully developed by Swid Powell. Opposite: Robert and Trix Haussmann at work in their Zurich office.

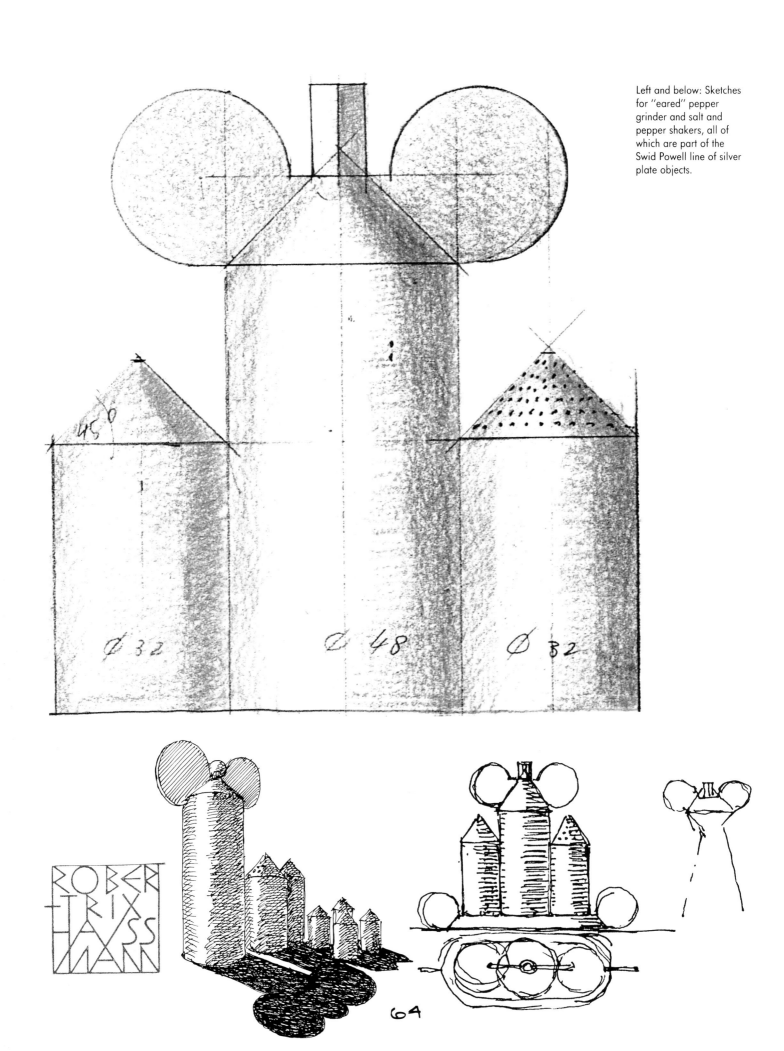

Left and below: Sketches for "eared" pepper grinder and salt and pepper shakers, all of which are part of the Swid Powell line of silver plate objects.

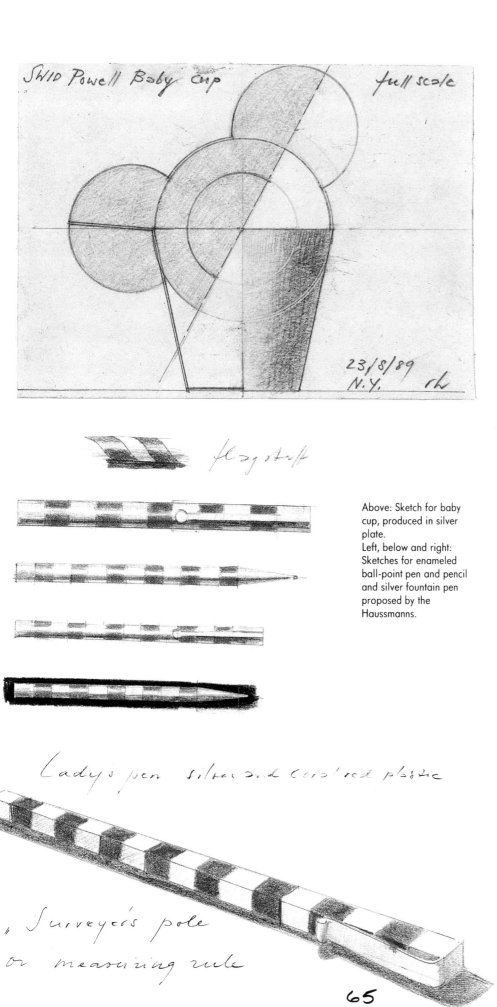

Above: Sketch for baby cup, produced in silver plate.
Left, below and right: Sketches for enameled ball-point pen and pencil and silver fountain pen proposed by the Haussmanns.

Above: Black-and-white *Stripes* pattern plate. A white-on-white version is also produced.
Opposite left: *Lehrstück IV "Seven Codes"* clothes wardrobe, designed by the Haussmanns in 1978. One can see the relationship between the architects' own work and their patterns for Swid Powell.
Opposite right: Studies for *Platonic* plate pattern.

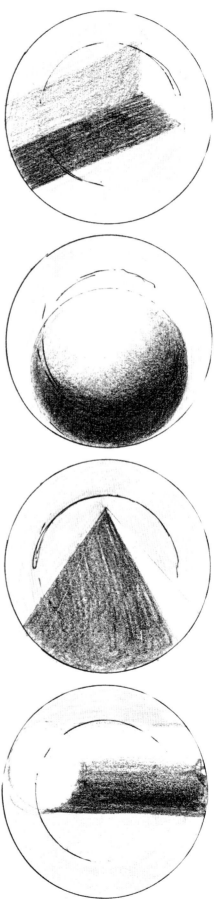

67

Steven Holl

"For an architect, life begins at forty-five," says master-builder Philip Johnson. At forty-two, Steven Holl has already distinguished himself as an innovative force in a new generation of American architects. In a world of "isms"—Modernism, Post-Modernism and Deconstuctivism—Holl insists that he is none of the above. "I don't have a style, but rather an approach that has to do with space, light and materials: the tactile dimensions of architecture," he says.

Holl operates on the principle of what he calls "limited concept." Because every site and its condition are different, there must be a unique approach for each situation. "I try to build a meaning into the site, so for me it's like starting over again on every project." Holl's portfolio includes a rustic board-and-batten house on Martha's Vineyard, a concrete and stucco commercial-residential building in Seaside, Florida, and the sleek Pace Collection shop in Manhattan, with its stark planes of colored glass and lines of black metal.

For Holl, the products he has designed for Swid Powell are closely linked to the architectural ideas he was developing at the time. "My plates—*Linear*, *Planar* and *Volumetric*—were an exploration in composition. This idea was related to the curriculum of a first-year architecture class I was teaching at Columbia, in which 'zero-ground' geometry was studied, involving problems in linear, planar and volumetric composition."

Holl's intensely intellectual approach found a worthy challenge when he turned his attention to designing candlesticks. "I was able to find a relationship to what I was thinking about at the time, which was to investigate the realm of material in which something is made. The idea was to take material, apply several coats of acid and watch a natural process pull out the color—in this case a raw patina green. Having done that, I considered the notion of space. I set the candlesticks high, so that when they're placed on a table, you see through them rather then look at them. We're not talking about just an object for the table here."

If Holl concentrated on form and matter, it was because he was quite familiar with work on this scale. "All my architecture, because of my idea of the tactile, gets down to the small scale all the time. I'm always thinking about light fixtures and door handles—about hardware. The larger scale has to have the smaller scale in it."

Opposite: *Planar* plate by Holl.
Right: Watercolor sketches for picture frame and oil can. Both were eventually produced in metal, but the oil can became a bud vase.

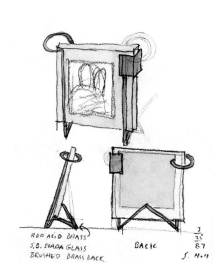
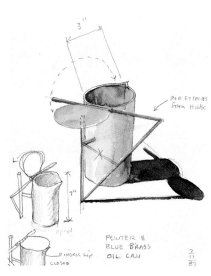

Arata Isozaki

If anyone exemplifies the changing nature of contemporary architecture, it is the Japanese architect Arata Isozaki. Throughout the 1980s, Isozaki has effectively blended the geometric clarity of Modernism with the traditional forms of Post-Modernism to create such high-profile commissions as the Palladium nightclub in New York City, the Museum of Contemporary Art in Los Angeles, a stadium for the Barcelona Olympics and the Tokyo Globe, a re-creation of Shakespeare's theater.

For Isozaki there is very little relationship between his architecture and his work for Swid Powell. "In my architectural work I don't rely on decorative elements, I use geometric form. The inspiration for my Swid Powell designs comes from my familiarity with the traditional patterns of Japanese porcelain that I've tried to apply to my plates."

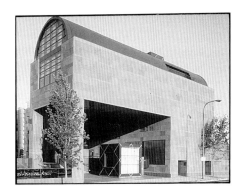

Opposite: Isozaki's *Stream* plate.
Right: The Museum of Contemporary Art (MOCA), Los Angeles, designed by Isozaki in 1987.

Richard Meier

In a time when converts fall away and rules are wantonly broken, one architect still holds to the tenets of the "true" faith. This is Richard Meier, and his is the creed of Modernism—a Modernism adapted and refined, but with Le Corbusier's passion for geometric clarity left intact. "If any architect under sixty represents, with undiluted conviction, the wish to preserve the classical legacy of twenties Modernism (but without its 'machine-for-living-in' rhetoric), that person is Meier," says *Time* art critic Robert Hughes.

The youngest recipient of the Pritzker Prize—architecture's equivalent of the Nobel—Meier has long been interested in the interplay of space, form and light. His signature color is white, which he uses to define these concepts and emphasize the sheer presence of buildings like the Atheneum, a visitor's center in New Harmony, Indiana; the High Museum in Atlanta; and the Museum for Decorative Arts in Frankfurt. He is now working on the largest architectural commission of our

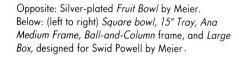

Opposite: Silver-plated *Fruit Bowl* by Meier. Below: (left to right) *Square bowl, 15" Tray, Ana Medium Frame, Ball-and-Column* frame, and *Large Box*, designed for Swid Powell by Meier.

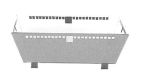 

Right: Sketch for a mantel clock. The bottom portion of the clock was developed as the *Candelabra,* produced in silver plate.
Below right: Studies for a series of silver-plated napkin rings incorporating the signature grid patterns Meier has applied to most of his Swid Powell product designs.
Opposite: New York architect Richard Meier at work.

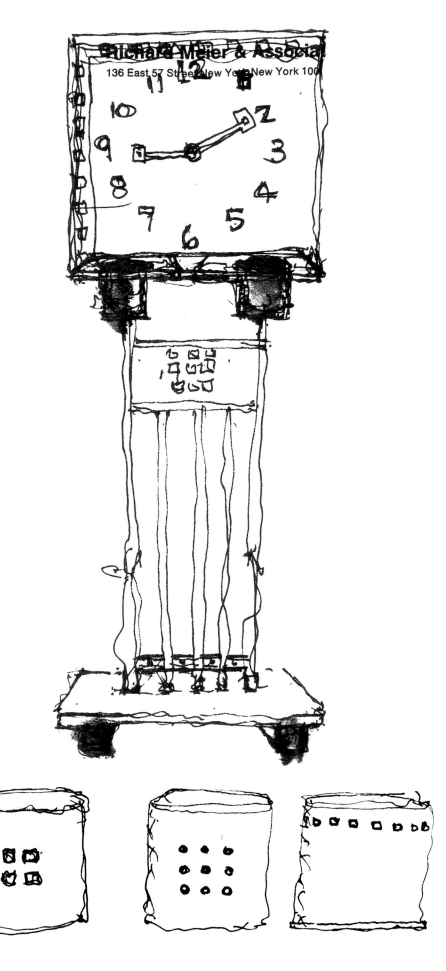

time, the $100 million J. Paul Getty Center for the Arts in Los Angeles.

For Meier, there is a straightforward relationship between his architectural work and the objects he has designed for Swid Powell. "I relate these products to my architecture in that they're related to me," he says. "What I design are not only things I want to live with, but things I need. For example, I need photo frames for pictures of my children, I need bowls for nuts. I also design for Swid Powell because I haven't been able to find things on the market that I like to live with on a daily basis. It satisfies my own particular desire for certain kinds of things that are part of life for me and for others who want them. It's the same as architecture."

Though Meier acknowledges that the similarities between his buildings and his silver pieces lie in formal concerns such as form and proportion, he is also quick to counter the idea of a more tangible correspondence. "In general terms, there is a relationship, but do my objects look like a building? No. Are they intended to look like a

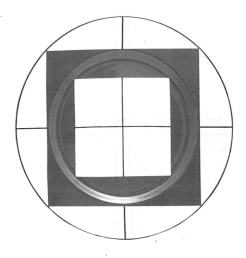 

building? No. They are intended to be what they are, which is of a scale and a use very different from architecture. The cut-out squares on my silver collection for Swid Powell do give it a scale, in the same way windows and doors lend a scale to buildings. But there is no direct correlation between the grids on the High Museum and those used on these products. The intention of the cut-out squares is to break up the surface, to be reflective and decorative. It's a different scaling device."

For Meier, it is the difference in scale that provides a satisfying supplement to architecture. "As an architect, one of the things that appeals to me is the length of time it takes to design products for Swid Powell," he says. "At the Getty Museum, it will be ten years before I see my labor. It's rewarding to design something, make a prototype and, within a year, see your design is realized."

His architecture seeks to achieve a true harmony of form and function, and Meier has the same goal for his Swid Powell designs. "Architects are practical people. When I design something, I say to myself, 'I want to use it, I want it to feel right, I want it to look right, and I want it to work.' I'm not designing a letter opener to sit on someone's table as an object, I'm designing it to open letters—and if it's not sharp enough to open letters, then I don't want it. At the same time, it also has to be something that, when you pick it up in your hand, it feels good. But in the end, if it doesn't work, it doesn't matter if it looks good."

Above left: *Joseph* buffet plate design.
Above: *Peachtree* buffet plate design.
Right: *Signature* buffet plate. Meier designed this pattern over lunch with Swid. He scrawled out two lines on a piece of paper and handed it to her. She was impressed with the simplicity and strength of the design, and put it into immediate production.
Opposite: Swid Powell's most successful silver piece, the *Meier Candlestick,* shown as produced and in sketch form.

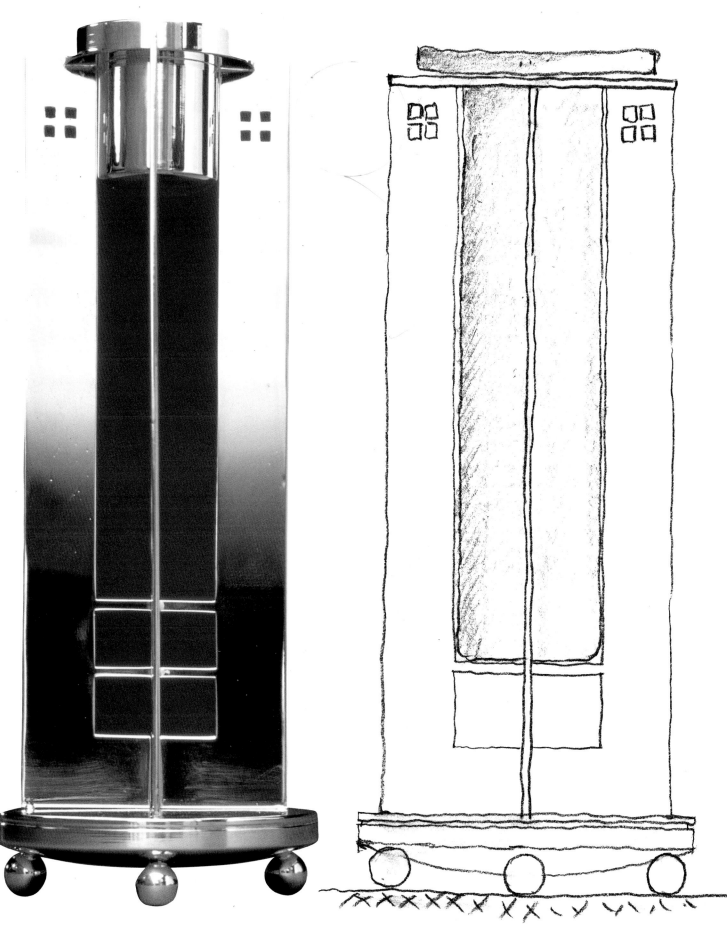

Vase 2 type 1

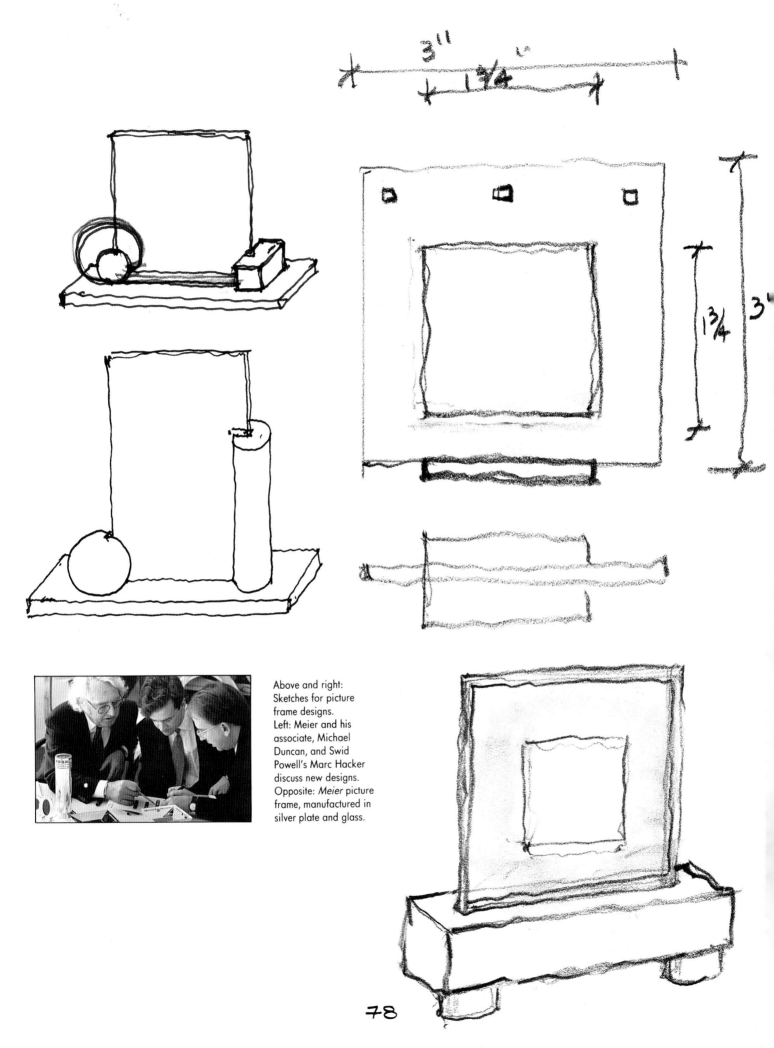

Above and right: Sketches for picture frame designs.
Left: Meier and his associate, Michael Duncan, and Swid Powell's Marc Hacker discuss new designs.
Opposite: *Meier* picture frame, manufactured in silver plate and glass.

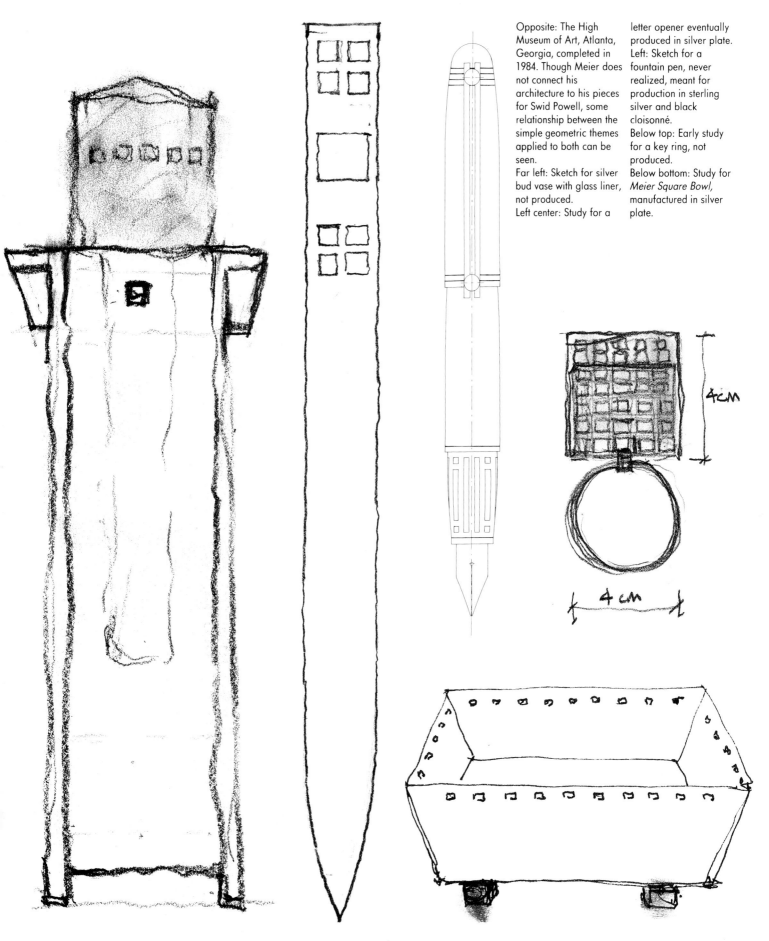

Opposite: The High Museum of Art, Atlanta, Georgia, completed in 1984. Though Meier does not connect his architecture to his pieces for Swid Powell, some relationship between the simple geometric themes applied to both can be seen.
Far left: Sketch for silver bud vase with glass liner, not produced.
Left center: Study for a letter opener eventually produced in silver plate.
Left: Sketch for a fountain pen, never realized, meant for production in sterling silver and black cloisonné.
Below top: Early study for a key ring, not produced.
Below bottom: Study for *Meier Square Bowl*, manufactured in silver plate.

David Palterer

Opposite: Palterer's sketches for candlesticks, produced in silver plate.
Below left and right: Drawings for plate designs, not produced.

David Palterer is an Israeli-born architect who works in Florence, where, as he remarks, "it's a common phenomenon for the worlds of architecture and design to meet." His recent projects include the restoration and expansion of the San Casciano Theater, the Florence Airport, a public park in Pesaro and a bird park in Tel Aviv.

For Palterer, architecture isn't about a given scale. "It is about the way one approaches 'designing.' Working on a small scale involves a much more immediate and precise relationship between invention and realization; between emotion and function," he argues.

The motif that runs through his architecture and his work for Swid Powell is "the compatibility between the rule and the exception, between geometric and organic." A candlestick, he says, is a "useless" object, but one with great emotional power. "It is an ideal center point, its light defining a space. The precarious and flickering luminosity of the flame that melts the candle makes visual the measure of time. The flame itself reminds me of the sun, of gold; it radiates energy and heat that is reflected in the pedestal of silver, the moon. The candlestick makes me think, with some nostalgia, of the time when science was alchemy. But then all objects I care about have a sense of *ceremonial* importance that I can't explain but which can perhaps be felt."

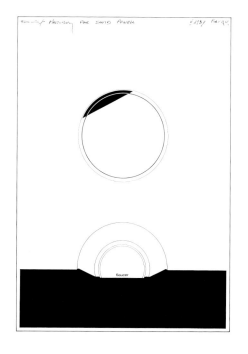
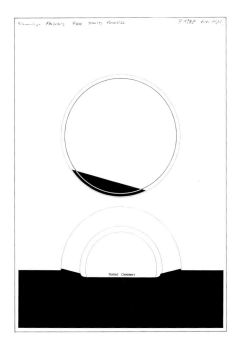

Paolo Portoghesi

Paolo Portoghesi, the Italian architect and teacher, is a Post-Modernist who has concentrated a great part of his critical and design career on the architectural history of the Italian Renaissance and Baroque.

He has built extensively in Europe over the last thirty years and has written numerous books on architecture. Since 1982 he has been a professor of history at the School of Architecture at the University of Rome; since 1983 he has been the director of the architecture division of the Venice Biennale.

"As an architect, I consider myself a pupil of Francesco Borromini, one of the masters of architectural details," says Portoghesi. "For this reason, I consider the details of a building very important—they can transmit to the observer a variety of messages that clarify the whole of the structure. After the observer has taken in the dimension of the site, he indulges himself by looking at the small details. For me, that's the perfect moment to communicate the secret and magic feeling that exists in every architectural adventure and in every object that is related to architecture."

For Portoghesi, there is a direct relationship between architecture and designing products for Swid Powell. "My buildings and objects share the same identity. The objects which embellish a house and which we use in everyday life are like architectural details; they give a feeling to the rooms in which we live. And that expands our means of communication. It is especially rewarding to design on a small scale because it gives a more intimate dimension to my work as an architect."

Opposite: *Eclipse* buffet and salad plates.
Below left and right: Studies for plate. Prototypes were made with this pattern but not developed further.

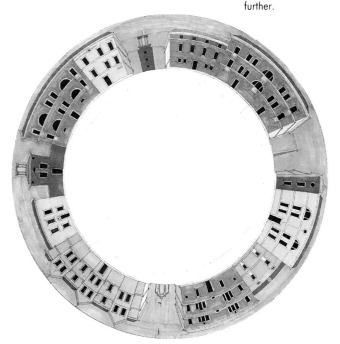 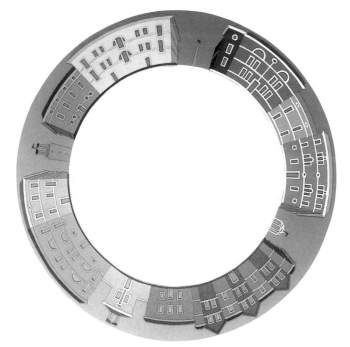

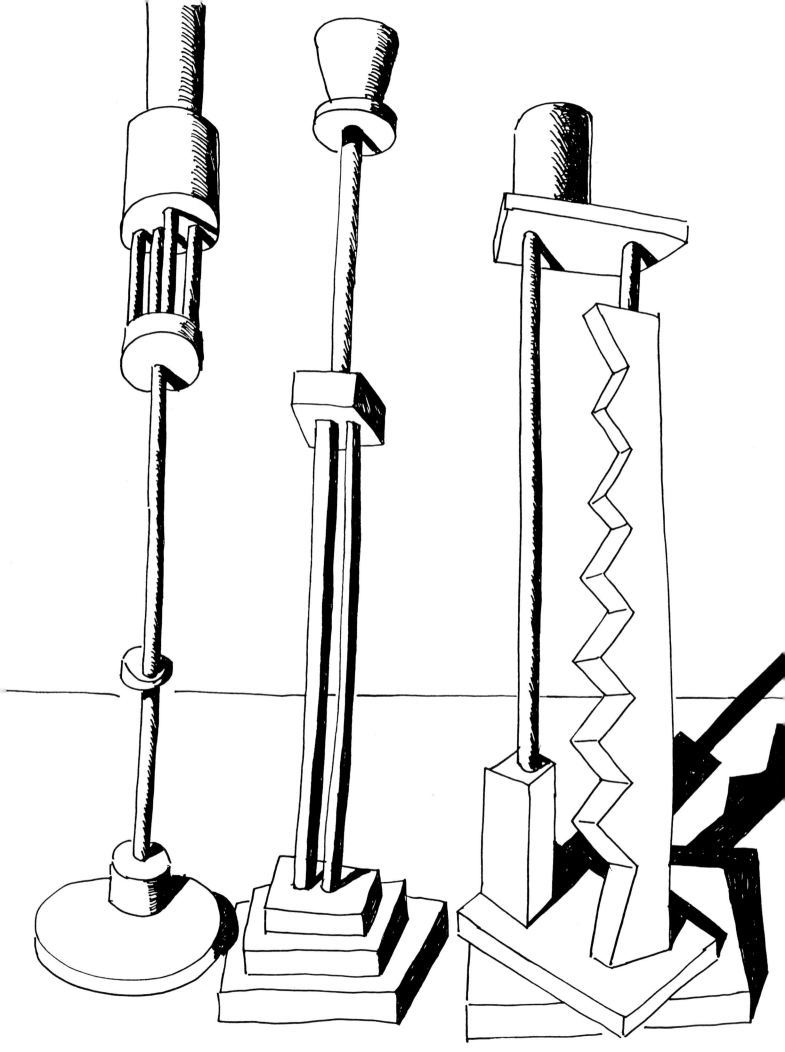

Ettore Sottsass

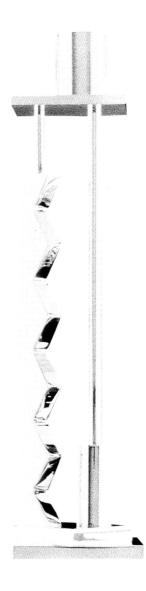

For nearly thirty years Ettore Sottsass was known as the Italian designer who created highly respected products for mainstream corporate clients like Olivetti. Then, in 1981, which might have been considered the twilight of his career, he broke out of his role as "industrial designer" and, with a group of young colleagues, launched one of the most revolutionary and influential design movements of the twentieth century. This movement—which he called Memphis in honor of a Bob Dylan song—showcased a collection of furniture and objects that became instantly recognizable. With its vibrant colors, bold patterns and jarring forms, Memphis was a violent reaction against the austerity and repetitiveness of Modern architect-designed furniture. A decade later, at age seventy-three, Ettore Sottsass has done another turn. He now focuses on architecture and interiors with such projects as the Esprit showrooms in Düsseldorf and Cologne, the Rainbow fabric company headquarters in Milan, and apartments in Tokyo, New York and Venice. In the San Juan Mountains of Colorado, he has recently completed the first house that he's built from the ground up.

His experience with Swid Powell marks the first time that Sottsass has designed for an American company. And though his tableware designs

Opposite: Sketches for (from left) *Starlight, Moonlight* and *Silvershade* candlesticks drawn by Sottsass.
Above: Silver- and gold-plated *Moonlight* candlestick.
Right: *Silvershade* candlestick, produced in silver plate.

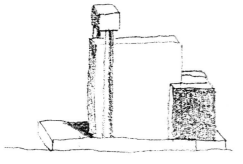 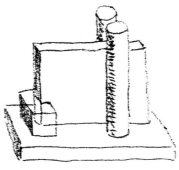

Top left: Sketch for a fruit bowl in marble and silver plate, not produced.
Top right: Sottsass and Swid meet to discuss designs.
Above left and right: Preliminary studies for picture frames in marble and glass.
Below: Sketch for a picture frame in marble and glass, not produced.
Opposite: Milan-based architect/designer Sottsass at work.

are a far cry from the flamboyant playfulness of Memphis, they share the same jarring edge. "Because Swid Powell was very much linked to the tradition of American design, I had to be careful not to seem like a fish out of water," explains Sottsass. "When I design, I always think, unavoidably, of the consumer, and in this case I tried to imagine myself as an American consumer. I thought of a sumptuously laid American table with candles. The only problem was, I never use candles."

It is, however, precisely in his Swid Powell candlesticks that Sottsass uses the same kind of figurative motifs he explored in the Memphis furniture. In his other work for Swid Powell, he draws upon decorative elements of the past and personal experience to create a style that is somewhat different from Memphis. "In the *Medici* china I wanted to make a design that used iconographic elements of Renaissance floral decorations," says Sottsass. "I didn't use these decorations in their entirety, but instead I made them explode into small bits and reassembled them in an order different from the original one."

As a result of his many trips to India in recent years, Sottsass became interested in Eastern

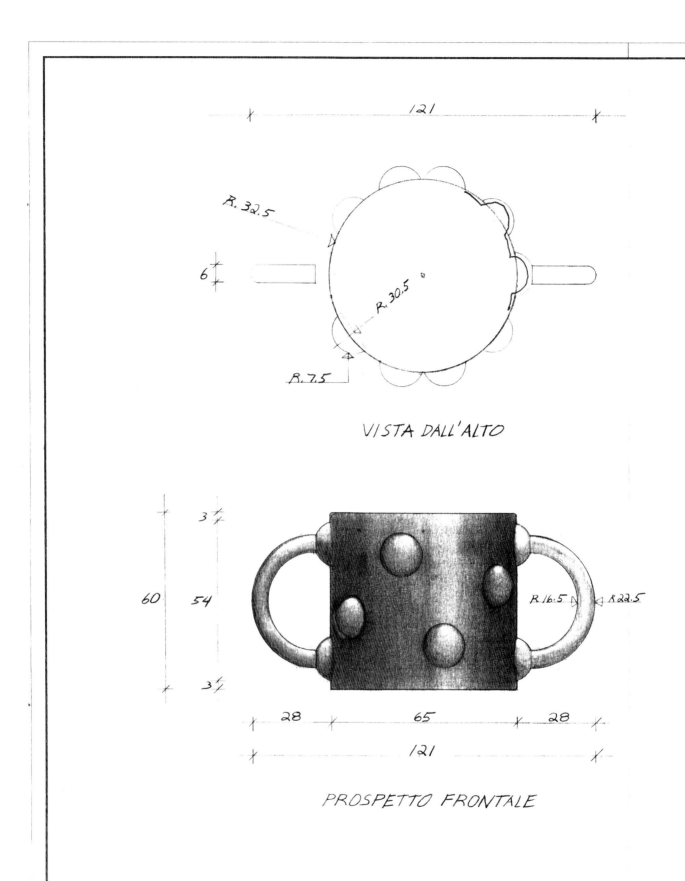

VISTA DALL'ALTO

PROSPETTO FRONTALE

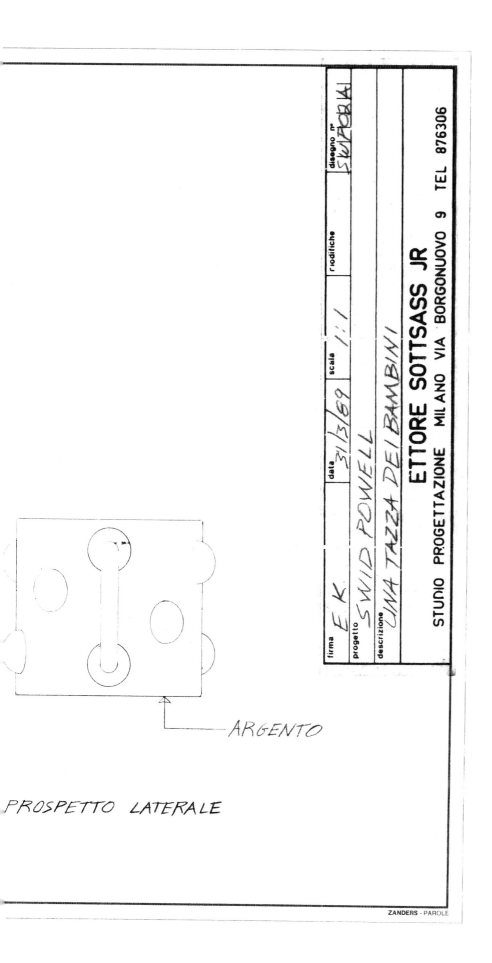

decorative arts. "The *Madras* has more or less the same idea as the *Medici* except that I was using Yantra iconography and, of course, different colors and proportions," says Sottsass.

For Sottsass the difference between architecture and product design is minimal. "Architecture can tackle more sophisticated issues," he says. "But it's the same philosophy. I believe in producing for people, not industry. In this commercial world there still can be spaces where people don't hide themselves, but find themselves."

And after more than forty years of designing everything from typewriters to computers to furniture and now tabletop objects for Swid Powell, how does Sottsass connect it all? "I cannot see how the work I've done for Swid Powell could not be related to my other work. But *how* they are related is something the critics should find out, not me," he replies.

Left: Working drawings for a silver-plated baby cup, now in production. Overleaf: Drawing, *Madras* pattern plate. Page 93: Studies for *Renaissance* pattern for plate and cup decals.

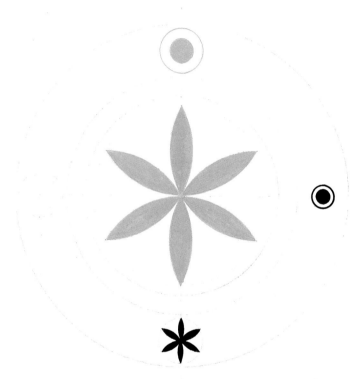

SALAD DESSERT

PROPOSAL (A)

SOTTSASS '86
FOR NAN SWID

George Sowden

In 1970, George Sowden abandoned what he calls "the nationalism of England" and emigrated to Milan for an industrial-design career with Olivetti. But a fascination with the decorative arts, coupled with the permissive atmosphere of the Radical Design movement in the early 1970s, led him to reconsider his choice of career. In 1980, his talent for pattern and color led him to become a force in the frenetic energy of the Memphis group.

Sowden now operates his own atelier in Milan, where he designs everything from interiors and furniture to rugs and clocks. Of his work, Sowden says: "When I design furniture or decoration, textiles or objects, I express a message or an idea with them. I look for a meaning, an essence or a communication. Sometimes it is frenzied, sometimes discordant, but mostly I look for something gentle, to create an experience or a sensation of pleasure; and through pleasure I aim at beauty."

For Sowden, design is a search, not a solution. "Working with Swid Powell," he says, "helped open doors leading to other worlds of design to explore . . . with more questions for me to answer."

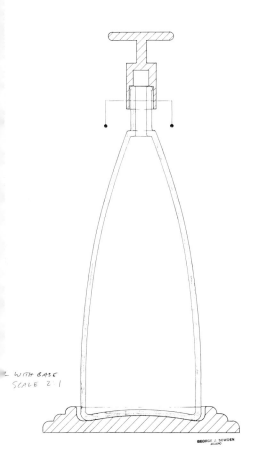

Opposite: *Rio* buffet plate and mug by Sowden for Swid Powell. Below left and right: Oil and Vinegar containers in silver plate and glass, not produced.

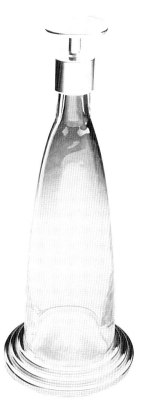

Robert Stern

Robert A. M. Stern stands alone in Swid Powell's stable—he is the only Post-Modernist who designs houses that would be familiar to a nineteenth-century client. He has advocated the belief that architects should look to the past for inspiration and instruction in his architecture as well as in his roles as professor of architecture at Columbia University, author (*New Directions in American Architecture* and *New York 1900*) and narrator of the PBS series "Pride of Place."

Stern probes the meaning of the past in the many shingled and gabled houses he has designed in Long Island's South Fork, Martha's Vineyard and in a number of suburban housing developments. He has also been responsible for such diverse projects as Congregation Kol Israel Synagogue in Brooklyn, the Prospect Point Office Building in La Jolla, California, and the Sprigg Lane Dormitories at the University of Virginia.

As he has to his architecture, Stern brought modern traditionalist ideas to Swid Powell. "An architect can bring a certain perception which he has from the scale of buildings

Opposite: Stern's *Moderne*-patterned plates.
Below: Drawings for a tea set, not produced.

Above top: The Bozzi House, East Hampton, New York, 1983, is an example of Stern's work in the vernacular of the Shingle Style.
Above: Sketch for *Triptych* picture frame.
Opposite: Stern in his New York office holding a *Moderne* cup, with *Century* candlesticks, prototypes for other candlestick designs and *Moderne* plates.

down to the scale of the interior of buildings and then beyond, to the things that furnish interiors," he explains. "Historically, there has always been a pleasure in the minaturization of certain architectural ideas, such as candlesticks that look like classical columns. One pair I did for Swid Powell—*Metropolitan*—is by no means unusual in that tradition; it's just unusual in the particular design. I love old things as I love old buildings, and I try to emulate that in all my work. But emulating the past is very different from buying something old, which is often the wrong scale for the present. For example, if you buy a nineteenth-century candlestick and put it in an apartment with eight-foot ceilings, it's usually out of scale and everything has to be restudied. As in my architecture, I try to reconceive these old ideas to fit into contemporary houses. For me, the pleasure is in reconceiving things from the past with some new things I've picked up along the way."

For Stern, the consideration of the small scale is a constant in all his work. "Every problem has its own scale," he says. "The bigger the project the more important it is to find small-scale elements to humanize it and to integrate it with a person's experience. In my work for Swid Powell, I'm considering the smaller scale that many people live in and the casual arrangements in contemporary living—even if we have what we call a formal house. People who live formally today don't hold a candle to the way people lived formally in the nineteenth century. So I'm trying to bring in the qualities of that day in a fresh new way."

Inspiration for Stern comes not only from architectural elements; it's also personal. "I designed the *Metropolitan* candlesticks because I have a pair of eighteenth-century candlesticks that I love, and my challenge was to see if I could design a pair that could sit comfortably next to the antique ones—which they do."

"My challenge is to make things that are 'today,' that can be made in our way and be affordable and can take their place in the world of new and old things. We all live with the new and the old. If you buy a dress or a suit, you expect it to be made as nicely as it can be, so you can walk into a room with antiques and not feel ridiculous. This division between the past and the present that artists and architects still rather preposterously argue for is what I try to prove false in everything I design."

Left: Stern and Swid surrounded by models and prototypes for numerous designs. Below: *Stern Fruit Bowl*, produced in silver plate. Opposite: The elegant proportions of Stern's silver-plated *Water Pitcher* make it a classic Swid Powell product. Overleaf top: Stern submitted these candlestick designs at the first design development meeting for Swid Powell. Overleaf bottom: Early designs for plates, not produced. Page 103: The *Century* candlestick.

CANDLESTICKS FOR SWID-POWELL   ROBERT A.M. STERN ARCHITECTS
JANUARY 5, 1983

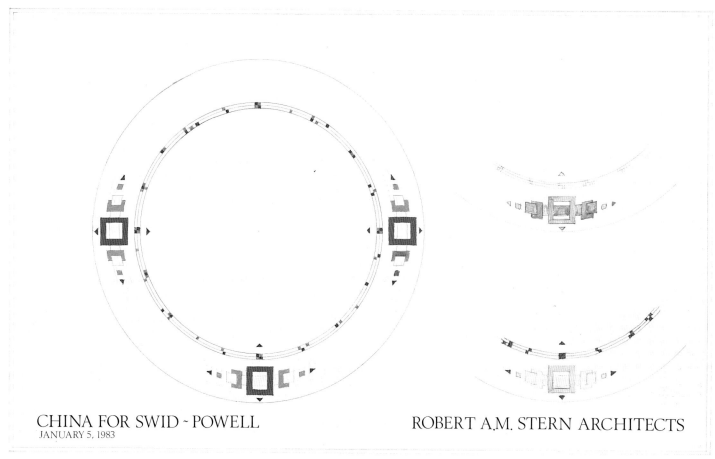

CHINA FOR SWID-POWELL   ROBERT A.M. STERN ARCHITECTS
JANUARY 5, 1983

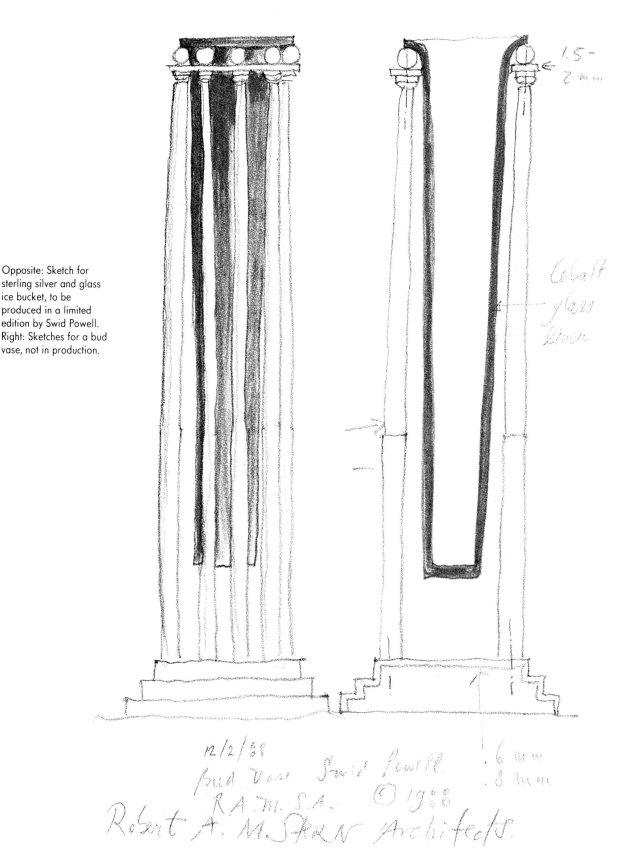

Opposite: Sketch for sterling silver and glass ice bucket, to be produced in a limited edition by Swid Powell. Right: Sketches for a bud vase, not in production.

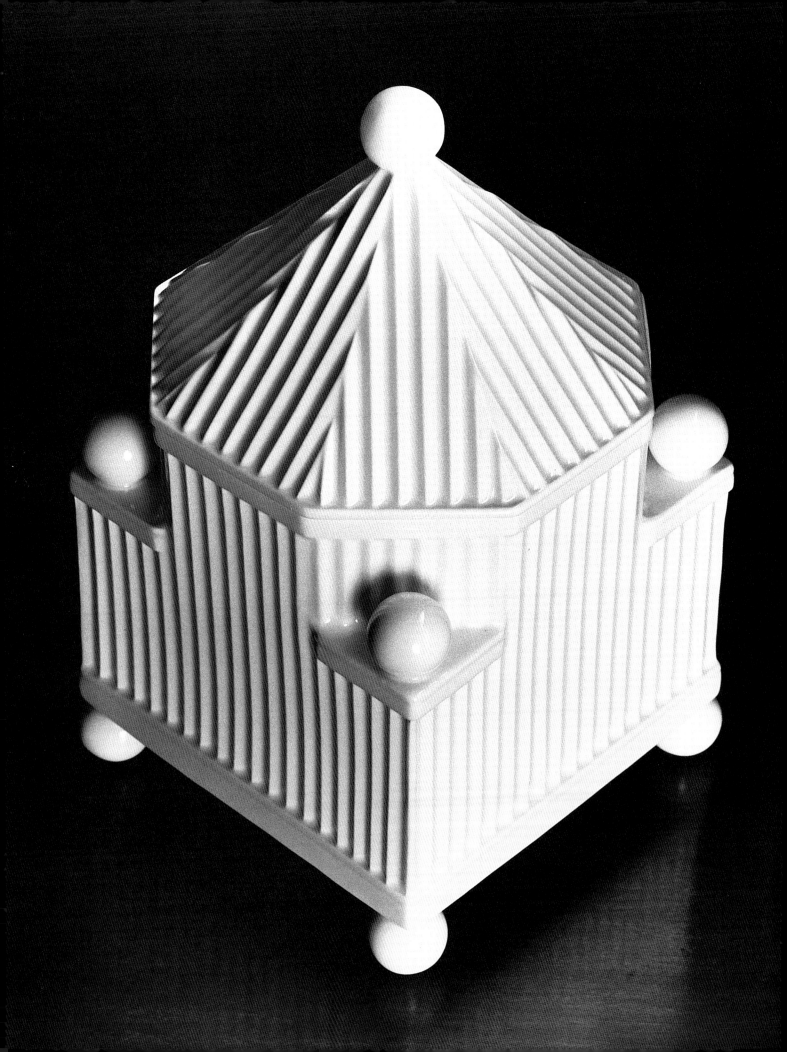

Stanley Tigerman

In a city steeped in the tradition of Mies van der Rohe and Frank Lloyd Wright, Chicago architect Stanley Tigerman has long been considered playful and irreverent, an inveterate maverick who refuses to be pigeonholed. For Tigerman, this has less to do with the intellectual debate between classicism and Modernism than it does with personal character.

"I've always rebelled against authoritarian forces," he explains. "If I can relate to anyone, it's Picasso, because he worked his way through something, finished it and went on to the next thing. With me, the client is key. I design for my clients, and they're all different. It's much easier to do signature work. I've made a hard life for myself. Every project is a new struggle for me."

With his wife and partner, Margaret McCurry, Tigerman has produced a portfolio that ranges

Opposite: *Tigerman Cookie Jar.*
Right: Studies for his and her tree houses, a commission sold in the 1985 Neiman-Marcus Christmas catalog.

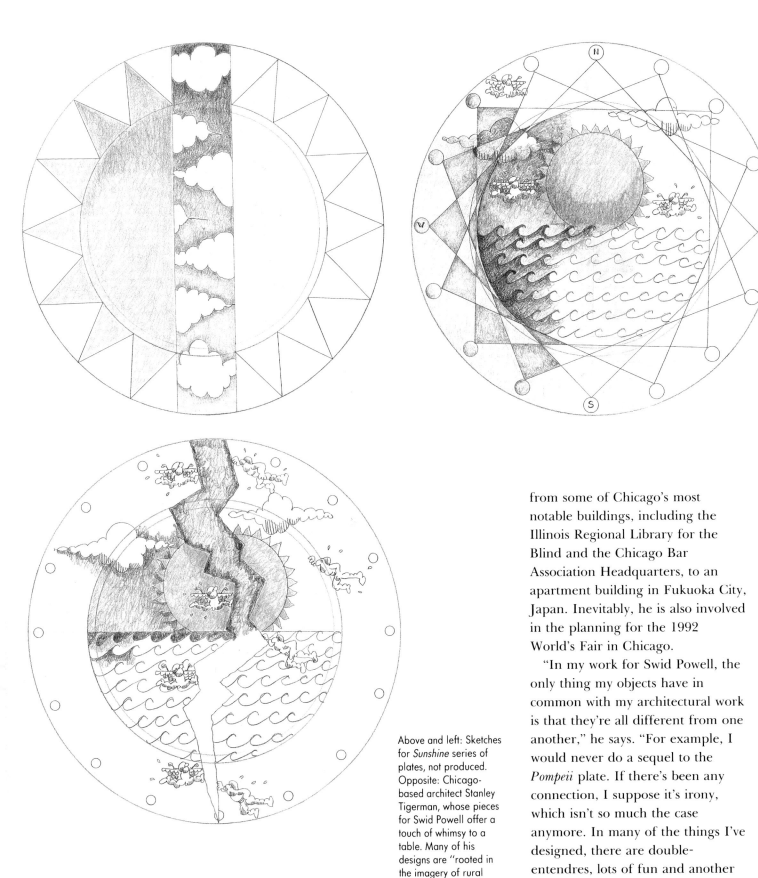

Above and left: Sketches for *Sunshine* series of plates, not produced. Opposite: Chicago-based architect Stanley Tigerman, whose pieces for Swid Powell offer a touch of whimsy to a table. Many of his designs are "rooted in the imagery of rural American life—home, family and farm," according to the architect.

from some of Chicago's most notable buildings, including the Illinois Regional Library for the Blind and the Chicago Bar Association Headquarters, to an apartment building in Fukuoka City, Japan. Inevitably, he is also involved in the planning for the 1992 World's Fair in Chicago.

"In my work for Swid Powell, the only thing my objects have in common with my architectural work is that they're all different from one another," he says. "For example, I would never do a sequel to the *Pompeii* plate. If there's been any connection, I suppose it's irony, which isn't so much the case anymore. In many of the things I've designed, there are double-entendres, lots of fun and another agenda. The *Verona* dinnerware relates to certain aspects of classical architecture. There is irony in the

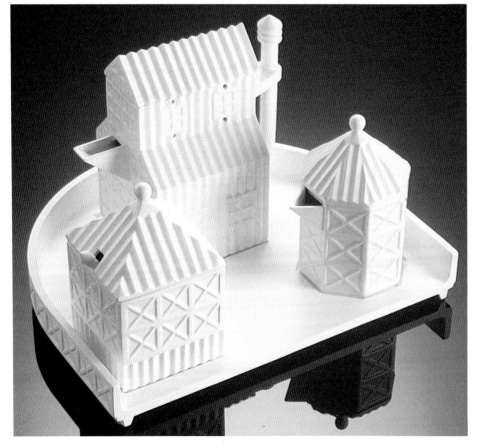

cup and saucer; the cup uses the same referent in elevation as the saucer, but you can rotate it and line up the columns."

Tigerman insists that he is still in many ways a child, and that his incessant sketching is the source of most of his decorative ideas. "I'm always doing little drawings of creatures, angels and soldiers," he explains, "and they wander through my work." His cookie jar is his version of a calliope, "a circus piece for kids." Even the one direct architectural reference in his Swid Powell collection is whimsical; his tea set is a re-creation of his weekend house in Michigan.

For Tigerman, the challenge of working on a small scale can be more daunting than designing buildings. "Anything you design takes a huge amount of time to do well," he explains. "And yet, with artifacts, the object can't appear overly wrought or overly worked. At the same time, it can't be flip. You can mask a lot of ineptitude in feet and acres; when you get down to inches and millimeters, it's hard to hide anything."

For all that, working with Swid Powell continues to offer an irresistible attraction. "I love serving dinner off plates I've designed in houses I've designed," says Tigerman. "It's a very old-fashioned architectural conceit that goes back to the time of Hadrian. Architects are insatiable people. When you're a designer, you want to put your hands on anything, on any scale. Everything is susceptible to the voraciousness."

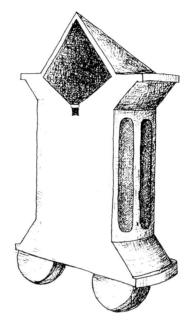

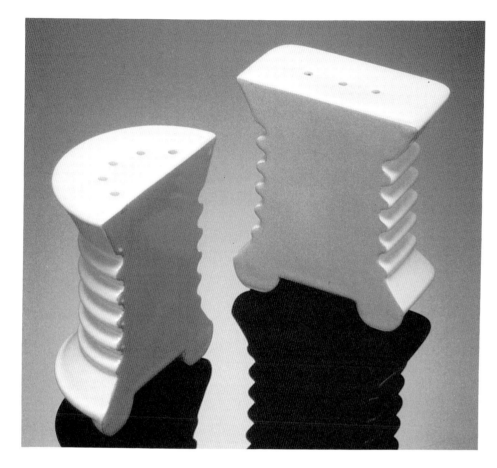

Opposite top: Architects' Weekend House, Lakeside, Michigan, 1983 (in association with Margaret I. McCurry). Opposite bottom: *Teaside* tea set by Tigerman for Swid Powell. The obvious model for this tea set is the architect's own compound pictured above.
Left: Sketches for a sugar bowl and creamer presented at an initial design review in 1983.
Above: *Tigerman Salt and Pepper Shakers.*
Right: Swid and Tigerman in discussion.
Overleaf: Sketches for candlestick series. Prototypes were made in metal, porcelain and marble, but the series was never fully developed.
Page 113: Rendering, *Pompeii* plate, in production.

Robert Venturi

With the publication of *Complexity and Contradiction in Architecture* in 1966, Robert Venturi became the enfant terrible of Modernism. One epigram was all it really took. Instead of adhering to the Modernist motto that "Less is more," Venturi defined himself—and a generation of new architects—with his proclamation that "Less is a bore." What architecture needed, he argued, was "messy vitality," with its myriad references to history, symbolism, form and popular culture. This was, as it happened, a viewpoint that the world of design was ready to adopt, and Venturi suddenly found himself the creator—and guru—of the Post-Modern movement.

Venturi has implemented his philosophy in such landmark buildings as his mother's house in Chestnut Hill, Pennsylvania, and the Guild House, a home for senior citizens in Philadelphia. With his partner and wife, Denise Scott

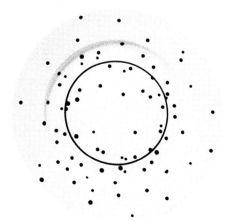

Opposite: *Notebook* cup and saucer.
Right: *Vegas* dinner plate by Venturi.

Brown, and partner John Rausch, Venturi has designed the Gordon Wu Hall and the Lewis Thomas Laboratory at Princeton University. Soon to be completed is the firm's extension of the National Gallery in London.

"There is definitely a connection between the buildings and the work for Swid Powell," Venturi says. "In the architecture I use applied ornament from many different sources. Or, as my partners and I say, 'We do buildings with Queen Anne in front and Mary Ann behind.' We like Pop elements and high art elements—as I do in my decorative art work. Another similarity is that I like to use historical reference in a way that is representational. I've done classical columns on buildings with a flat, two-dimensional representation; I've also done this in my candlestick for Swid Powell, where a Chippendale design was used and made three-dimensional."

At the opposite end of the spectrum, Venturi has incorporated his affection for the ordinary in the tableware that he has designed for Swid Powell. "In both *Grandmother* and *Notebook*, I wanted a universal and, at the same time, American motif—which is another way of saying I wanted to do something that was ordinary-pretty. Once again, I'm making it representational. *Grandmother* was a material source and *Notebook* a paper source. By putting these motifs on a hard edge, they become less strict and not overly sentimental."

But for all the correlation, Venturi says, "the source for these products is even more general. Quite simply, I love decorative arts." This love of the decorative can be

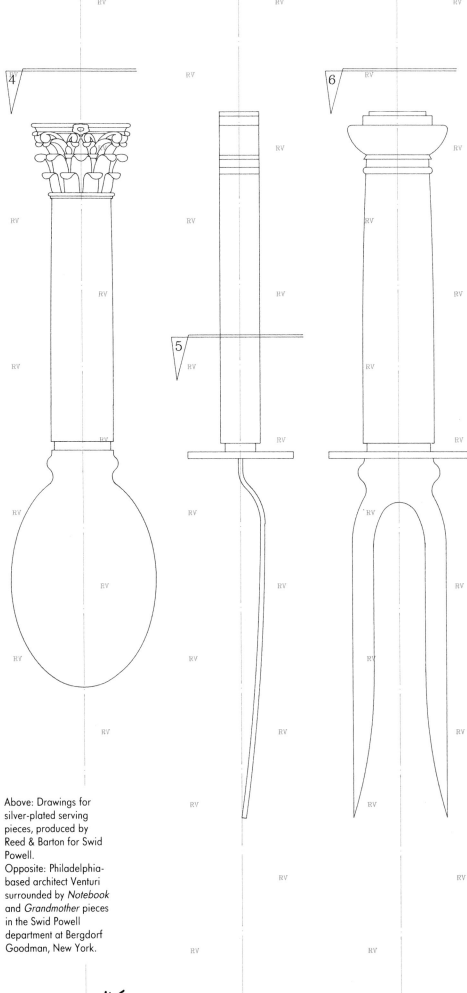

Above: Drawings for silver-plated serving pieces, produced by Reed & Barton for Swid Powell.
Opposite: Philadelphia-based architect Venturi surrounded by *Notebook* and *Grandmother* pieces in the Swid Powell department at Bergdorf Goodman, New York.

frustrating. "Because architecture is my first concern and responsibility, I have to do the decorative arts on the side. And as an architect, there's the problem of working on a small scale. To jump from architecture to these objects, where a sixteenth of an inch or a thirty-second of an inch literally means a lot, is difficult."

Venturi takes consolation in a historical perspective. For him, designing for Swid Powell is reminiscent of the Rococo or Art Nouveau periods, when interiors and architecture were connected to the furniture and the beauty came from the unity. But Venturi would never subscribe to a conventional historical view. "Our approach is, of course, more eclectic, our sources of form and symbolism more diverse. I glory in the richness of effect rather than in the unity of effect."

Top left: Sketch for *Village* teapot, in production.
Above left: Sketch for teapot, never developed.
Left: Grand's Restaurant, Philadelphia, 1967. This coffee-shop facade, with its stylized coffee cup-shaped sign, is an example of Venturi's "celebration of the mundane."
Opposite: Sketch for *Village* coffeepot.
Opposite right: Powell, Venturi and Swid celebrate the architect's work at a Bergdorf Goodman event.

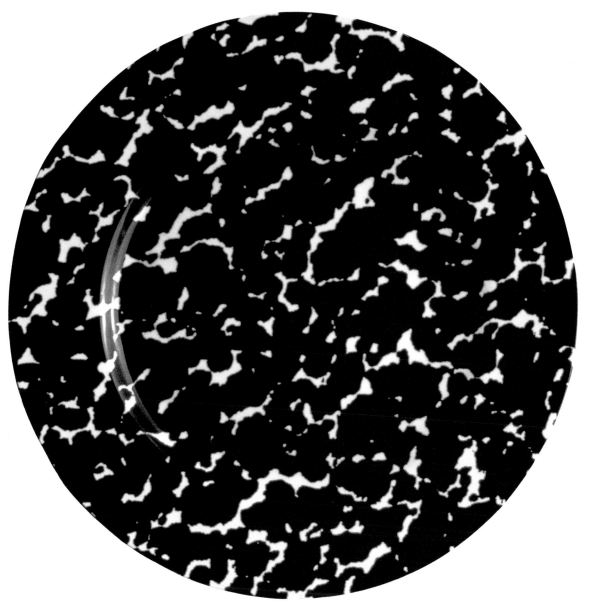

Opposite: The *Grandmother* pattern was inspired by a tablecloth design Venturi saw at the house of a colleague's grandmother.
Left: The *Notebook* plate has a pattern from an obvious source.
Below: The *Village* coffee and tea set consists of a coffeepot with the shape of a Tuscan tower, a Pantheon-shaped teapot, a peasant's hut sugar bowl and a *palazzo*-shaped creamer.

Left and opposite: Sketches for the *Venturi Candlestick,* produced in silver plate.
Below top: *Dashes,* a double-old-fashioned glass.
Below bottom: *Grandmother,* a Venturi glass design.

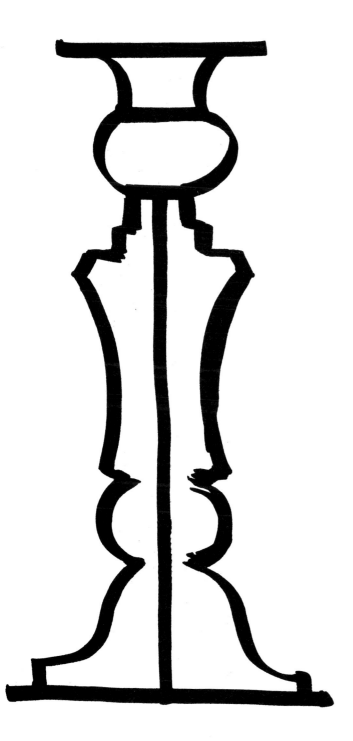
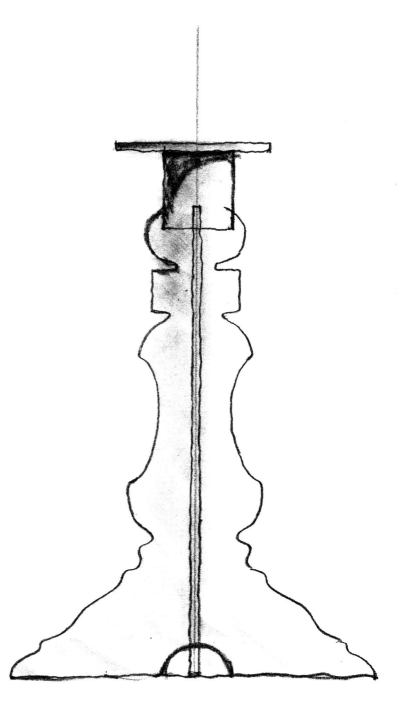

123

The Design Process

As Stanley Tigerman has wryly observed, architects don't like to be accused of being mere functionalists. "The issue is *transcendence*, taking things beyond what they portend to be," he says. "And that, happily, is also the way Nan Swid and Addie Powell look at it. They understand that architects are different from industrial designers. There are secondary and tertiary readings of buildings and objects that architects seek because of their training. An industrial designer may design something that is ostensibly more beautiful at first blush, but with certain exceptions his product won't last as long as one done by an architect. For an architect, it's not just a single shot."

Transcendence is an exalted idea. Production is a gritty reality. For Swid Powell, the initial—and continuing—challenge is to take the ideas that seem so promising in an architect's drawings and deliver them to the consumer at a high

Opposite: The *Tuxedo* coffeepot on the wheel at one of Swid Powell's Japanese porcelain factories. Below: Gwathmey Siegel sketches for the *Tuxedo* coffeepot.

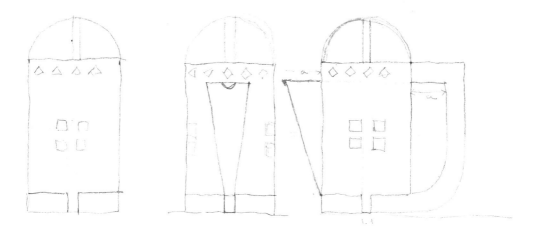

125

Right: The *Melting Candelabra* by the New York firm SITE proved to be too expensive to mass produce. The prototype was made of silver-plated cast bronze.
Below: *Nouveau* sugar bowl and creamer by Venturi was prototyped but not produced.
Opposite top: SITE's *Moon* pattern plate, not produced.
Opposite bottom: Influential architect Peter Eisenman designed the *Daedalus* plate in textured, sandblasted glass. Pattern only made it to the prototype stage. A companion plate in porcelain, *Icarus,* was produced in a limited edition for the Wexner Center for the Visual Arts, The Ohio State University, Columbus, Ohio, in 1989.

level of quality, functionality and affordability. These challenges were so daunting that, until Swid Powell came along, innovative products for the home were created only by industrial designers.

Fortunately for Swid and Powell, they were joined at the start by Marc Hacker, who coordinates their production efforts. "Marc Hacker is a quiet force who caused my things to be better than I was capable of making them," says Tigerman. For Gwathmey, Hacker is something more: "a phenomenal co-designer and collaborator."

An extremely modest man, Hacker is more likely to recount the difficulties of mass producing architecturally designed products than he is to share anecdotes about collaborating with the field's biggest names. "Most of these architects were designing products for the home for the first time. Like me, they had no experience in manufacturing silverplate or porcelain. While I was learning the disciplines of those materials from scratch, they were designing in a vacuum—often without keeping in mind the materials they would be working in. The problem was they were designing what they wanted, they weren't designing simple products. And that compounded the difficulty."

The architects were very ambitious. They wished, in fact, to reinvent the wheel, or rather, the plate. "People were trying to invent complicated things, like the first ashtray that flew to the moon," Stern says. "I told Nan, 'You can't afford these, it would be a miracle if fifteen people wanted them. Why don't we just paint and design plates?'"

Left and opposite: Painted patterns with color specifications and plans for Holl's *Volumetric* (top) and *Planar* (bottom) plates. The decal manufacturers follow these detailed charts to produce the desired effect.

problems is color," explains Hacker. "You silk-screen the pattern on to a decal sheet which holds the paint together. Soaked in water, the decal is transferred onto the plate; the plate is then fired in the kiln and the pattern marries with the glaze. The problem is that different colors come to life at different temperatures, and to get the color combination right can take many tries. The Steven Holl plates, for example, had eleven different colors being fired at once. It was very difficult to achieve the correct balance of color. And for some of their pieces, Michael Graves and Gwathmey Siegel came up with their own palettes and had factories mix new colors, which called for additional rounds of test firings."

If the plates were difficult, working in silverplate was even trickier. The Meier candlestick was an instant classic that still sells strongly, but when Meier first designed it, it was an intricate object, very difficult to make, and sure to be unaffordable for most consumers.

"We went through three or four prototypes, each one made by hand, before Richard signed off on it," Hacker says. "But none of us—including the manufacturer—focused much on the production technique. And there is a big difference between making something individually and making it in a production run. In this particular case, the manufacturer didn't understand the ins and outs of production. The first run arrived in New York wrapped in straw, not even individually packed. The silver balls fell off. It was a good lesson in

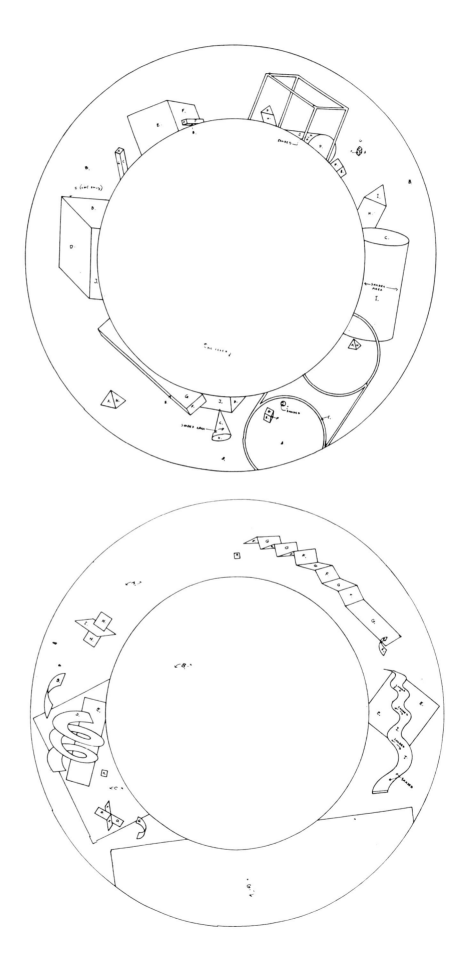

Above: Marc Hacker, Vice President of Design and Development for Swid Powell.
Right: Model, Tigerman's *Wings* bowl at a Florence, Italy, factory. Prior to making molds, porcelain manufacturers make a full-scale model of the object to be produced. A mold is cast around this model.

the importance of thinking *everything* through."

By learning to be extremely precise in its instructions to manufacturers, Swid Powell was able to cut its development time from eighteen months to as few as six. And the architects learned to design from a production standpoint. "When I'm working with the architects now, I'm able to give them much more information about the consequences of design decisions," Hacker says. "This isn't to prevent them from following their creative path, but it does give me the opportunity to say, 'If you were to do it this way, we could make this product for a more reasonable price.' It turns out that our architects are very interested in producing things for less. They're problem solvers—it's as if they have a client, a budget and a site. And that mindset makes them an increasingly fascinating design resource, because they are not creating in a void, vagrantly floating in the ether of design. They're actually thinking, 'How does it hold a candle? Will it fall over?' In other words, it's now a two-way process."

This does not deter Swid Powell from commissioning designs that have built-in production obstacles. Each Holl candlestick took five hours to make—the acid application had to be done by hand. Here, the retail cost reflects not the raw material, but the labor. Meier's hand-blown crystal stemware was expensive to engrave because the design called for Meier's signature grid, and a square-cut is difficult to execute.

The Holl candlesticks and the Meier glassware made it to the marketplace. Not everything has—

Above: Steel stamp for the manufacturing of the *Venturi Candlestick*. Opposite: This factory in Vicenza, Italy, is a metal-stamping and silver-plating operation where many Swid Powell objects are made. The stamping machine shown is used to make boxes.

and not for reasons that anyone at Swid Powell could have anticipated. James Wines of the New York firm SITE designed a plate that was a photograph of the craters of the moon. "It's a fantastic piece," Hacker says. "We previewed it to buyers at our trade show and it didn't catch on, so we never put it into production. And that's happened to every architect. Sometimes an element of design turns out to be fashionable or seasonal, and if a designer uses a color that is completely out that year, it can kill the product. Like Javier Bellosillo, the Spanish architect who designed the *Ruin* plate with colors that are reminiscent of old Santa Fe. If that had come out when the American West design look became popular, it would have sold far better than it did."

As the 1980s drew to a close, Swid Powell's success came to be measured in a new way—just as they had used architects as their design resource, large companies approached them to undertake a whole series of new products. Now Swid Powell has licensing agreements with Cannon/Fieldcrest, and Venturi, Tigerman, Sottsass, Meier and Graves are designing sheets that will serve the bedroom as their tableware does the dining room. With Reed & Barton, four Swid Powell architects—Meier, Graves, Venturi and Palterer—are preparing sets of serving pieces. And in 1989, Swid Powell was commissioned to create a new line of dinnerware for American Airlines' international first and business classes. The firm selected Charles Gwathmey to design the dinnerware.

This level of acceptance means that Swid Powell and its designers can work more quickly to bring new and more challenging products to market. By the late 1980s, Swid Powell branched out to include the work of fine artists in the tableware line. In 1989, the firm introduced plates with photographs of flowers by Robert Mapplethorpe. These were images familiar to buyers, and it was crucial to reproduce them at the same level of quality as the original photographs. It took five or six prototypes to get the right silk-screens for the decal. More recently, Swid Powell has been working with New York painter Donald Sultan on dinnerware.

The greatest achievement, however, is the increased ease of the collaborative process. "Some architects—Meier, Venturi, Graves and Stern—produce a complete set

Above: A handle is applied to a Gwathmey Siegel *Tuxedo* coffeepot. Opposite: (top to bottom) Serving pieces from the Swid Powell collection produced by Reed & Barton: Ebony and stainless steel cake set by Palterer; carving set in silver plate and stainless steel by Robert Venturi; salad servers of stainless steel and black cloisonné by Meier; silver-plated serving spoon and fork by Graves.

of drawings," says Hacker. "You can go directly to the factory with these drawings and make a prototype. With others there is a critical dialogue. The architect produces a sketch embodying an idea for a product, which we work on to produce a set of drawings. We then go back with those drawings to the architects. Recently, Stanley Tigerman sat down and made sketches for a winged platter, sugar bowl, creamer and baby-photo frame. From these I did full-scale sketches, then I met with him and he suggested ways to change them. We'll go back and forth until we go to prototype. It's a much smoother process than it was at the beginning. Sometimes things do get easier in this business."

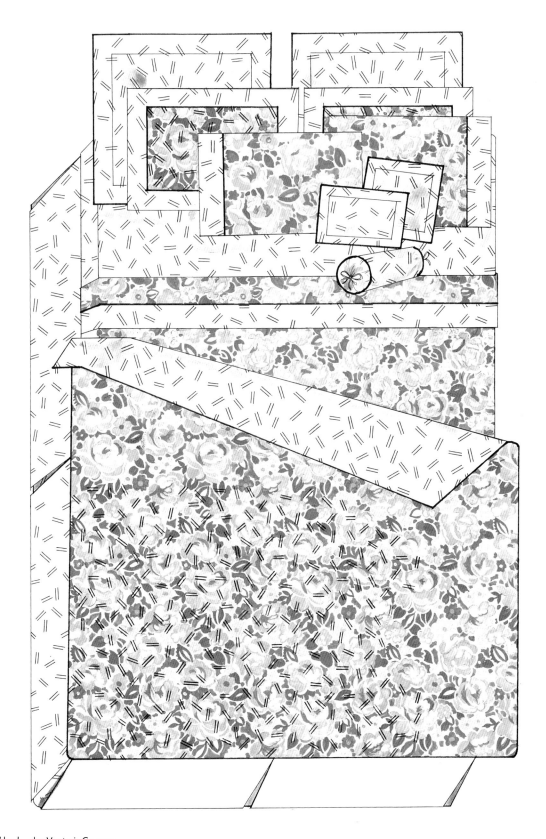

Opposite: *Metropol* bed linens designed by Meier for Swid Powell, manufactured by Cannon Mills.
Above: Bed sketch for *Grandmother* bed linens by Venturi. Cannon manufactures sheets, pillowcases, comforters, shams and blankets under license from Swid Powell.

This highly refined production process is not visible in the finished product. And that is crucial to the success of the products. The energy expended to make the object drops away, leaving only the realization of the architect's intention. For Hacker, that moment is key. "The interesting and fun thing about this business is that you think it's all going to be predictable, and then it's not, even with the products you think are relatively easy to make. There is always some issue like a tooling cost you didn't expect or a strike at the factory. That's why my greatest pleasure is in realizing the design. The greatest reward is when successful prototypes arrive in the office."

It is only when the finished prototype is put into full production that the months of development—discussion, creation and refinement—come to an end. At that point, it is the buyers who are the final arbiters—who seal the fate of the object.

Opposite: At a Florence factory, a metalworker solders a pitcher designed by Meier.
Top right: Passengers in American Airlines' International Business Class are served on Swid Powell dinnerware designed by Gwathmey Siegel.
Right: Gwathmey Siegel-designed china is part of the First Class cabin experience during international flights on American Airlines.

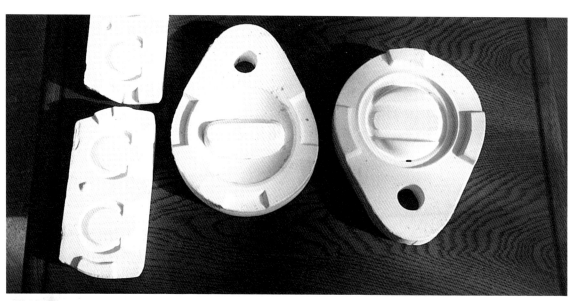

Left and below: Ceramic giftware in the Swid Powell line is made in these molds. Opposite: Approximately half of Swid Powell's porcelain production takes place in Japan. Here, some products dry on carts in a Japanese factory.

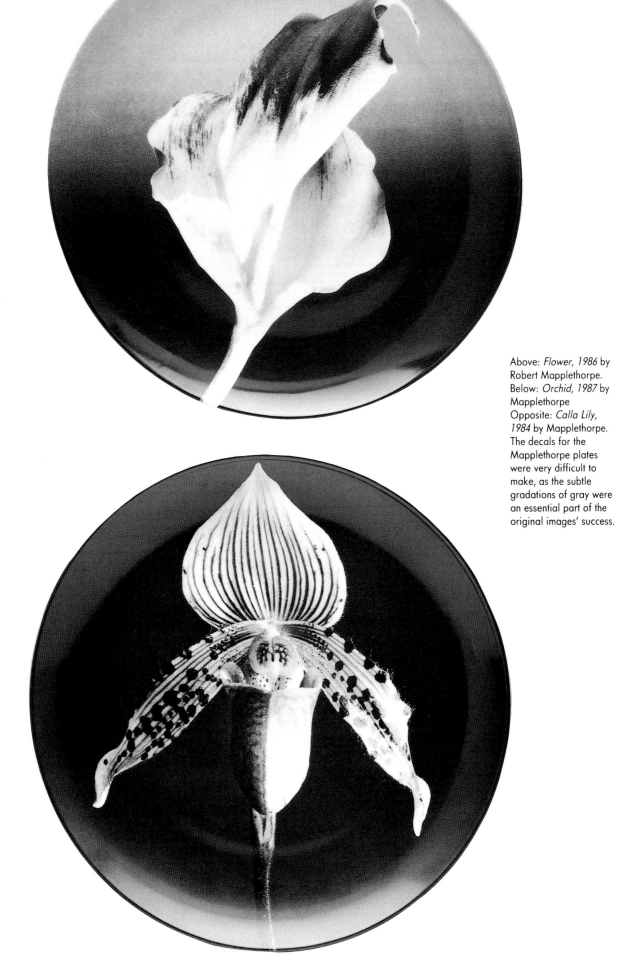

Above: *Flower, 1986* by Robert Mapplethorpe.
Below: *Orchid, 1987* by Mapplethorpe
Opposite: *Calla Lily, 1984* by Mapplethorpe. The decals for the Mapplethorpe plates were very difficult to make, as the subtle gradations of gray were an essential part of the original images' success.

PL 13424
$24.00
NIBS